DATE DUE

Principles of Contaminant Hydrogeology

Christopher M. Palmer

With Contributions by

Jeffrey L. Peterson
Jerold Behnke

LEWIS PUBLISHERS
Boca Raton   Ann Arbor   London   Tokyo

Library of Congress Cataloging-in-Publication Data

Palmer, Christopher M.
　　Principles of contaminant hydrogeology / Christopher M. Palmer :
　with contributions by Jeffrey L. Peterson and Jerold Behnke.
　　　P.　cm.
　　Includes bibliographical references and index.
　　1. Water, Underground — Pollution.　2. Hydrogeology.　I.
　Peterson, Jeffrey L.　II. Behnke, Jerold.　III. Title.
　　TD426.P35　1991
　　628.1′68 − dc20　　　91-24703
　　ISBN 0-87371-280-3

Note: Methods and problem solutions are those commonly employed
　　　　by consultants today. This text is not meant to be a substitute
　　　　for rigorous groundwater hydrogeology training. Additional
　　　　technical information regarding groundwater contamination
　　　　investigations is presented in federal and state guidance
　　　　documents.

LEWIS PUBLISHING, INC.
121 South Main Street, Chelsea, Michigan　48118

Printed in the United States of America　　　3 4 5 6 7 8 9 0

Malo mori suam foe oami.

Acknowledgments

As with any effort of this type, many people assisted in the preparation of this book. Jeff Peterson is a knowledgeable hydrogeologist, and his help is appreciated. Thanks are also due to Jerold Behnke, professor of hydrogeology in the Department of Geological and Physical Sciences at California State University, Chico, who aided very much in editing the text. The comments of the Lewis Publishers, Inc., reviewer were very welcome for critical review and content. Review and comment by Jon Elliott helped to distill the morass of environmental regulations into a general introduction. Additional thanks are due to Gene Corriden, who suggested writing the book, an outgrowth of the groundwater monitoring class I taught for the University of California Extension. Kelly Johnson and Michelle Gordon typed and formatted the manuscript and index. Finally, thanks are due to all consultants, geologists, and hydrogeologists with whom we have worked and discussed the various subjects contained in the text. Much of the text material contains approaches used by consultants in everyday applied contaminant hydrogeology problem-solving.

Preface

This book is intended to be an introduction for newcomers to basic principles of consulting hydrogeology. Much of the material presented here has grown out of an introductory groundwater monitoring course taught by one author (Palmer) for the University of California Extension, Santa Cruz. Many students taking the course were eager to become involved in groundwater issues and cleanup, yet few had any idea of how subsurface investigations are performed. Also, many were unaware of investigative techniques in the unsaturated and saturated zones to comply with regulations for site cleanup. Hence, this text presents and reviews problem-solving approaches commonly used by consulting geologists and hydrogeologists.

An accurate assessment of a client's subsurface contamination problem within the context of applicable regulations is usually the goal of the consulting hydrogeologist's investigation. Accuracy of investigation data will have profound effects upon defining the contaminant extent and site remediation effectiveness and cost. Consequently, all investigation efforts should be expended toward understanding the subsurface processes that affect remediation, and gaining site closure.

Emphasis is placed on the basic geologic nature of contaminant hydrogeologic investigations. The hydrogeology may then be understood within the geologic framework; that is, where are aquifers and aquitards, water quality, porosity, permeability, and so on. Davis (1987) points out rightly that the field of hydrogeology occurs within geologic science, not an engineering or other field. The geologic investigation is too often simplified or not adequately understood, which can bring grief to a consultant. This is not to say that other disciplines of engineering, chemistry, and laboratory analysis are not needed or should be minimized. A contamination investigation often is a multidiscipline effort. However, individuals who practice "hydrogeology" should have broad academic training in geology and *extensive field* experience in performing subsurface data collection and interpretation studies. In this way, interpretation of data and contaminant movement becomes more comprehensible.

Although this book introduces general aspects of an investigation, the need to understand site geology, stratigraphy, and aquifer properties is the main goal. Often inadequately defined subsurface geology results in a lack of subsurface understanding, including knowledge of groundwater and contaminant movement. Typically, projects have limiting budgets and resources with which to study a contamination problem. There will never be enough money and exploratory boreholes drilled for an absolute "final" answer to all subsurface questions. Answers may fall into ranges due to the variability of the natural materials. The quantity and quality of subsurface site information collected by the consultant forms the basis for successful problem definition. Judgments of data adequacy must be made in every subsurface investigation, since each investigation is unique.

Various interrelated aspects to investigation execution, such as field logistics, drilling techniques, sampling protocols, and general concepts of contaminant movement and remediation, are presented. Professionals involved in this work should keep abreast of the literature, new equipment and analysis techniques, textbooks on groundwater and related subjects, and rapidly changing regulatory rules. The newcomer hydrogeologic consultant or regulator overseeing contaminant hydrogeologic projects often needs to weave different threads of the study so that the report information makes sense. Although the technical project objectives must be met, the legal, financial, and ethical objectives in consulting are as important. This book is neither a "magic bullet" with all the answers nor a "cookbook" approach for success in hydrogeologic consulting. Rather, it seeks to introduce newcomer hydrogeologists to the basic information gathering so that sufficient data are available to erect the required geologic, hydrogeologic, and contaminant models. If anything, experience in subsurface studies shows the importance of data extrapolation. The information collected should be the best attainable, given the time, money, and regulatory constraints, so that the data extrapolation will not be questioned as unreasonable.

The consultant must also deal with project budgets, client interaction, and regulatory requirements. When a client signs a contract with the consultant, it is usually to perform a specific scope of work — typically an investigation to solve the client's problem. However, this means that the geologist or hydrogeologist must understand the applicable regulations and their legal implications. The project budget has been proposed to cover a certain work scope, and the amount of information collected will be a direct function of the available money. A vexing problem that can arise is how much information is collected for the time and money expended? The investigation being performed may be reconnaissance or preliminary, and formal comprehensive reports may require several investigation efforts before the problem is deemed sufficiently complete.

Often the consulting hydrogeologist produces technical reports that will be read by a nontechnical audience. Information contained in reports should present the immediate problem, in clear terms supported by the site data, even when it may be bad news. The consultant should realize that his or her work may be critically reviewed years later by other consultants or regulating agencies. The information gathered in the subsurface investigation must be defensible and "state-of-the-art."

We hope that this book may assist both new and old consultants, and generally interested individuals, by outlining consulting approaches to consulting contaminant hydrogeologic projects.

Christopher Palmer holds B.A. and M.A. degrees in Geology from California State University, Fresno. He is a Registered Geologist, Environmental Assessor, and Well Inspector in the State of California, and has 11 years of experience in the execution and management of hydrogeologic and applied engineering geologic projects. The author or coauthor of several technical articles, Mr. Palmer has been lecturer and guest lecturer for classes at the University of California, Berkeley, and at Santa Cruz (Extension), and serves on the technical advisory committee at the University of California Santa Cruz. Mr. Palmer is presently Senior Geological Manager, supervising all groundwater contamination and aquifer remediation and geotechnical studies at Ensco Environmental Services, Inc., in Fremont, California.

Table of Contents

CHAPTER 7

CHAPTER 8

Principles
of
Contaminant
Hydrogeology

CHAPTER 1

Geologic Frameworks for Contaminant Hydrogeology Investigations

INTRODUCTION

Subsurface investigations into the presence and extent of contaminated groundwater are primarily geological investigations. The site subsurface geology forms the physical framework through which groundwater (and contaminant fluids) flow. Understanding the site geology provides the fundamental basis for understanding site hydrogeology and defining contaminant movement. This information will be of paramount importance when preparing models of contaminant transport and fate, and ultimate site remediation plans. The investigations also integrate aspects of soil engineering, applied chemistry, and environmental engineering disciplines. Although some surface geophysical techniques may aid in gathering subsurface data, site information must be collected by exploratory drilling and soil, sediment, or rock sampling. The geologic data are then put together to construct the site subsurface hydrogeologic environment.

Geologic environments will vary depending on where geologists happen to find themselves, but basic hydrogeologic questions (for example, depth to water, aquifer contacts) need to be answered. Indeed, this geologic and hydrogeologic information, as well as the subsurface sampling, is required by the regulating agencies in all cases.

Site investigations are conducted to get the following information on both the unsaturated (vadose) zone and saturated zone. Current regulations and cleanup are directed to both the groundwater and the area overlying groundwater that may have an adverse affect on the aquifer. Thus, the investigation should address the vadose soil and sediment, processes operational therein, and the geological potential pathways to the aquifer. Additionally, the investigation must retrieve information on the general site geology and hydrogeology, including

1

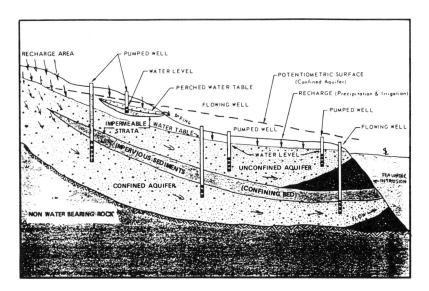

Figure 1.   Unconfined and confined water and hydrologic cycle (Cal. Bull. 118, 1974).

aquifer geometry, groundwater occurrence, flow direction and gradient, aquifer matrix, and groundwater physical and chemical sampling and aquifer testing. This provides the essential data on contact relations of aquifer and aquitard, potentiometric surface, and direction and speed of flow; identifies physical and chemical properties of the matrix; and stresses the system by pumping tests. This information is needed at the minimum in order to comprehend geology, groundwater movement, and aquifer properties. Since the contaminants occur and move in this environment, acquiring correct site data is vital to tracking contaminant movement and forming conceptual remediation models. After all, the primary function of the consulting geologist and hydrogeologist for this type of investigation is to identify, track, and clean up the pollution problem.

GROUNDWATER OCCURRENCE AND GEOLOGY OF AQUIFERS

Groundwater occurs in subsurface rock and strata called *aquifers,* which are comprised of porous and permeable material (see Figure 1). The aquifers are usually bounded by relatively "impermeable" bodies called *aquitards,* which do not readily transmit water. Aquifers may, of course, be composed of alluvium, crystalline, or sedimentary rock. Throughout this book, the aquifers will be discussed in a stratigraphic sense; that is, products of sediment deposition and sand as aquifer, and silt and clay as aquitard (or that the units of interest are aquifer and aquitard by the geologic nature of their composition). Also aquifers and aquitards will be conceptually presented as tabular bodies — a strong generalization, but used for ease of illustration for the subsurface exploration and movement of water and contaminants. When case histories deal with specific geology or groundwater flow conditions, the text will so state.

UNSATURATED AND SATURATED GEOLOGIC ENVIRONMENTS—A BRIEF REVIEW

Groundwater occurs in almost every type of geology, in both urban and nonurbanized settings. An exhaustive review of all geologic and hydrogeologic environments is beyond the scope of this book. Groundwater occurrence has been described in the standard texts, and the reader is referred to those for detailed general groundwater occurrence (see, for example, Davis and DeWeist, 1966; Freeze and Cherry, 1979; Heath, 1982; Fetter, 1988). A review of the unsaturated zone will be presented first, followed by a review of general concepts of groundwater occurrence and flow (groundwater hydraulics are presented in Chapter 8).

Unsaturated (Vadose) Zone

The unsaturated (vadose) zone is the region that overlies the saturated zone or, for our purpose, any aquifer (see Figures 2 and 3). If there is a contaminant release, the contaminants almost always pass through this region to get to the aquifer. Although this region is not saturated (all available pore space filled with fluid), areas may be locally saturated, and elsewhere fluid moves under a response to surface tension, gravitation, and capillary forces. Also, recent soil vapor testing and measuring techniques have been developed so vapor occurrence and movement can be recorded.

The vadose zone is geologically a very heterogeneous region. Typically soil formation occurs here, as well as sequential burial from fluvial, colluvial, and alluvial sedimentation. Since unsaturated zones

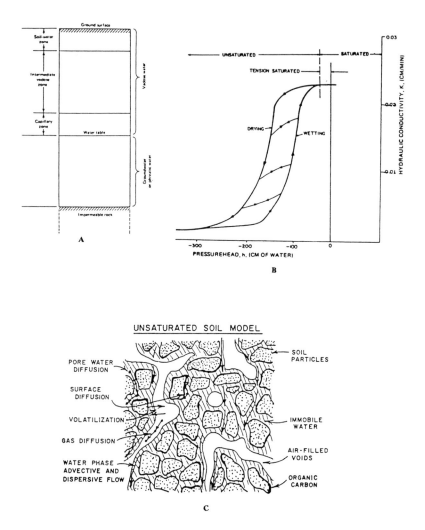

Figure 2.   Vadose Zone. A. Division of vadose zone and groundwater zone. B. Hysteresis effect of wetting and drying soil showing increased matrix suction with drying (after Freeze and Cherry, 1979 in EPA, 1985). C. Unsaturated soil model showing mass transport and retardation mechanisms (after Gierke and others, 1986).

may be hundreds of feet thick, the term *soil* is a catchall term to describe this material and is somewhat geologically misleading. For example, the site of interest may be soil-covered fractured rock, which contains groundwater below an unsaturated region including

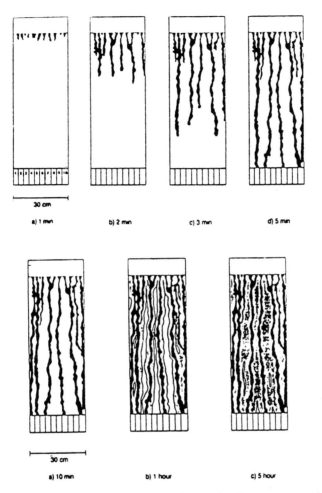

30 cm

a) 1 min   b) 2 min   c) 3 min   d) 5 min

30 cm

a) 10 min   b) 1 hour   c) 5 hour

Figure 3. Macropore flow experiment showing wetting front movement in vadose zone. Rapid vertical infiltration occurs when saturation occurs in larger innerconnected pores, followed by lateral wetting (Morrison, 1989).

soil, colluvium, and rock. A series of soils (or paleosols) may be buried sequentially, thus resulting in a crudely horizontally bedded deposit many feet thick, below the organic-rich soil formation region of 3 to 10 ft within the surface. The range in particle sizes, bedding, layering, presence of buried structures, and so on may have profound effects upon both unsaturated, saturated, and vapor movement (Morrison, 1989). Accurately ascertaining the geologic makeup of the vadose zone and its effect upon contaminant migration and movement can be very difficult and requires a substantial subsurface investigation effort.

Fluid movement in unsaturated conditions may be governed by one or more processes, depending upon the moisture content and local conditions. These four processes are hysteresis, macropore flow, capillarity, and saturated (or Darcian) flow.

Hysteresis

Hysteresis occurs when the unsaturated region is not devoid of moisture, but rather has fluid (water for our discussion) adhered to soil particles and grains. Flow between grains (pore throats) moves through micropores. A general equation for unsaturated flow (Fetter, 1988) is

$$\phi = \psi(\theta) + Z$$

where ϕ = total potential, unsaturated flow
$\psi(\theta)$ = moisture potential, measured as suction
Z = elevation head

Fluid movement is governed in response to local pressure gradients. The suction (usually measured in centibars per cubic centimeter) is the effect of a negative pressure head of soil and water on the unsaturated hydraulic conductivity (Hillel, 1980; Morrison, 1989) (see Figure 2). As the soil becomes saturated, the soil pressure, or suction, declines to zero as the available porosity becomes totally filled with water. As the soil drains, the pressure head increases and suction increases, with the remaining water held by tension. The change in moisture occurs constantly, so the process occurs as a matter of the changing moisture states depending on recharge of groundwater or contaminants. For contaminant studies, the most important difference between saturated and unsaturated flow is that the unsaturated hydraulic conductivity (K) is not a constant for different soil moisture contents (Morrison, 1989).

Macropore Flow

Macropore flow occurs when surface water enters a crack, or *macropore,* and literally flows into the crack. This vertical movement continues until the crack fills and local saturated conditions arise migrating from the crack (Figure 3). Openings and channels are common in soil and buried sediment from biologic processes by animals and roots, and from backfilled conduits and foundations. Desiccation cracks may be open to several feet, or deeper, and may be filled in with porous sediment, which would enhance flow and allow rapid and deep penetration of fluid through an otherwise "impermeable" layer, with the obvious potential movement toward the aquifer. Lateral saturation from the vertical flow would also occur, although at a slower rate and following saturation of the macropores (Morrison, 1989). This concept is exceedingly important since logging of "soils" with conventional logging techniques may miss these structures.

Capillary Movement

Capillarity occurs due to available porosity immediately above the saturated zone as a response to water surface tension allowing vertical movement against gravity. The height of the capillary fringe above the saturated area is dependent upon grain size. Hence, the height of capillarity into a clay or silt is higher than that of a sand or gravel. Recent data (API, 1989) indicate that the capillary fringe may be up to several feet above stabilized water levels in overlying clay soils and sediments above shallow aquifers.

Saturated Flow

Saturated flow may occur where sufficient water collects upon an impermeable layer to saturate the porosity, resulting in a "perched" groundwater lens above the aquifer itself. These perched lenses can be very extensive and may form miniaquifers contained within the greater unsaturated realm. Thus, Darcian flow can occur and may be used to model the fluid flow. Large quantities of water or contaminants may move vertically, then laterally, transmitting fluid to the aquifer over circuitous and not easily recognizable paths. Additional general mathematical concepts of saturated groundwater flow will be discussed in detail in Chapter 8.

Saturated Geologic Environments—Aquifers

A brief review of geologic environments encountered in consulting work may be classified into three very general types of environments: igneous-metamorphic rock, sedimentary rock, and alluvial filled basins. These three general types of environments are used for the following discussions for ease in text examples. Groundwater occurrence in the following discussions is generalized and somewhat similar to the models developed by the U.S. Environmental Protection Agency (EPA) for use in contaminant investigations, which use similar summaries (EPA, 1987a). The following summary is not an attempt to modify established hydrogeologic convention; the reader is referred to those references from which the following discussion borrows heavily. The intention is to introduce general groundwater occurrence in terms of porosity and permeability for different terrains. Obviously, the individual site needs to be evaluated for the specific geology and hydrogeologic characteristics. Figures 4 and 5 present diagrams and some ranges of porosity and permeability of rocks and sediment.

Igneous-Metamorphic Rock

Crystalline igneous or metamorphic rocks, and some very well lithified or slightly metamorphosed sedimentary rocks, comprise these terrains. Usually these rocks are "dry" in a large-quantity water resource sense, yield little groundwater to wells, and so are considered impermeable. Groundwater flow is typically through joints, cracks, or structural discontinuities or tectonic fracturing, with very limited porosity between mineral or metasediment grains. Where the fractures are dense and recharge is abundant, locally high water yields may occur. In igneous and metamorphic terrains, fractures tend to be common in the uppermost 300 ft, which will contain and yield most groundwater. Although some fracture systems may be very widespread and deep, typically the fractures close with depth, causing a decrease in groundwater yield. While laminar flow assumptions may not be valid in fracture flow, it may not be a major hindrance to describing flow and transport in fractured systems (Schmelling and Ross, 1988).

Sedimentary Rock

Sedimentary rocks usually contain lithified and semilithified sediments, and chemical sediments such as limestone. Porosity and permeability in sedimentary rocks tend to be greater than in igneous or metamorphic rocks. Groundwater resource occurrence is typically more abundant in sandstone and limestone formations, and limited in siltstones and shaley formations. Water flow is usually around grains and voids, although fracture flow may occur and could be a predomi-

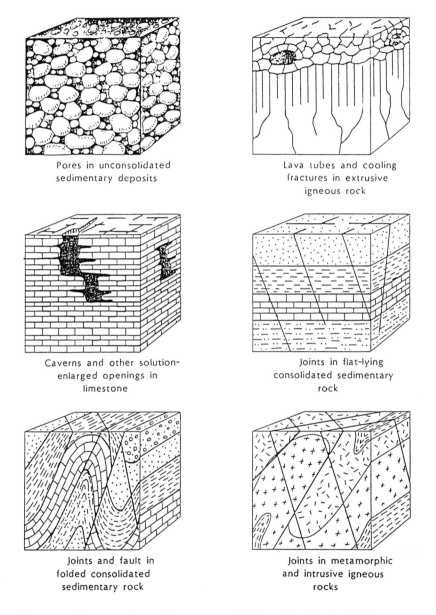

Pores in unconsolidated
sedimentary deposits

Lava tubes and cooling
fractures in extrusive
igneous rock

Caverns and other solution-
enlarged openings in
limestone

Joints in flat-lying
consolidated sedimentary
rock

Joints and fault in
folded consolidated
sedimentary rock

Joints in metamorphic
and intrusive igneous
rocks

Figure 4.   Intergranular and fracture porosity examples (Heath, 1989)

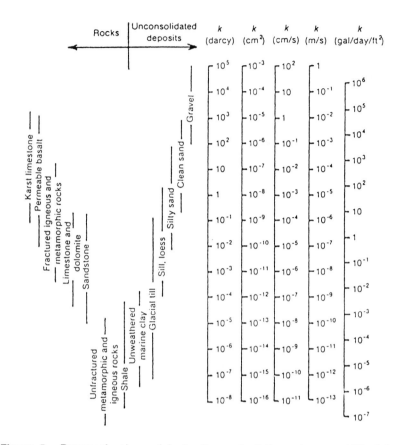

Figure 5. Range of values of hydraulic conductivity and permeability (after Freeze and Cherry, 1979).

nant flow path in well-lithified or cemented rocks. The "conventional" layered aquifer and aquitard systems can occur in these terrains, where the tabular aquifer geometry is usually envisioned.

Carbonate rock flow paths may range from flow through fractures and around grains to flow through caverns. Flow situations may become somewhat unpredictable since the solution voids, caves, and breccias may affect direction and speed of water movement. The recharge by streams may flow directly into voids and exit through other voids further downgradient. Tracers may be used to track flow in carbonate systems.

Alluvial-Filled Basins

Alluvial-filled basins are shallow or deep basins filled with alluvium deposited by fluvial processes and form large usable groundwater resources in intermountain areas in the western United States and in portions of the eastern United States. For example, the San Joaquin Valley (California) overlies thick sediment sections containing an enormous unconfined and confined aquifer system with tremendous groundwater reserves. Other basins may be small with limited groundwater reserves. The occurrence of groundwater is similar to sedimentary rock terrains: sands tend to form the aquifer, and clays and silts form aquitards in layered systems. Since larger urbanized areas tend to occur in flat alluvial-filled valleys, groundwater in these basins is more developed and may be more threatened than other terrains. In urban areas, the hydrogeologist often deals with alluvial, glacial, or colluvial aquifer occurrence.

Strata in alluvial basins occur in certain depositional environments that will dictate the geometry of the occurrence of sand and clay bodies (Reineck and Singh, 1986) (see Figure 6). The sediment may occur as strata continuous over large areas in certain environments; however, these strata can be either continuous or discontinuous on scales of yards or miles. Also, internal stratification causes texture gradations and interbedding within the overall sand or clay strata. Glacial sedimentation may result in clayey lake sediments, or tills containing variable clast sizes and sand beds. Hence, the aquifers and aquitards texture and strata correlation may change abruptly, and alluvial situations can be very difficult to work in given their changeable nature.

Perched Groundwater

A special hydrogeologic condition is created when perched groundwater is encountered. A perched condition arises when a quantity of water collects on an impermeable stratum above what would be locally considered the regionally extensive aquifer. These perched aquifers are usually limited in extent and quantity of water, which is recharged by either natural or human sources; however, these aquifers may become sufficiently large to be locally recognized water producers (see Figure 1).

Perched aquifers may form in numerous ways, and the following three examples illustrate some cases.

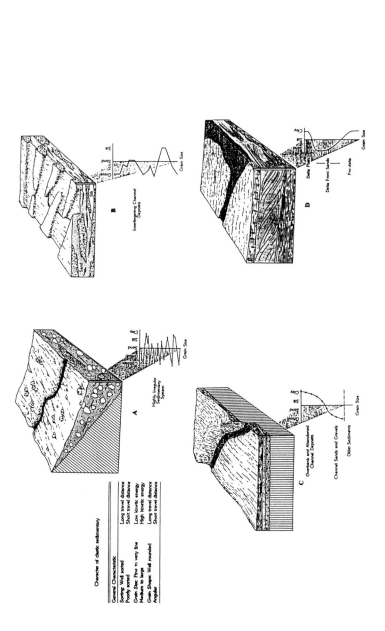

Figure 6. Alluvial deposits. A. Alluvial fan. B. Braided channel. C. Overbank and abandoned channels. D. Deltaic channel and plain. Vertical sediment packages differ given the type of deposit, allowing interpretation and subsurface correlation of strata (Mathewson, 1979).

1. Clay strata may be interbedded with sandy strata and form the impermeable layer upon which the water collects.
2. Alluvium or colluvium that overlies relatively impermeable bedrock may collect water at and above the sediment-rock contact.
3. Water may collect in interconnected fractures or joints of bedrock, but vertical movement is restricted by closing fractures at depth. Hence, the water tends to flow "downhill" in the fractures and is perched in the otherwise impermeable bedrock.

In many instances, the perched water, if present, is usually the first impacted by descending contaminants.

BRIEF REVIEW OF GROUNDWATER MOVEMENT

The following is a very brief review of groundwater movement and some of the general relationships of water flow commonly used in consulting work. This is not a substitute for referring to well-known texts (for example, Freeze and Cherry, 1979; Heath, 1982; Fetter, 1988) or for completing rigorous university groundwater or hydrogeology classes.

Subsurface hydrogeology of any area is typically divided into the unsaturated (vadose) zone and the saturated zone. Each zone is comprised of sediment or rock that possesses porosity, or the ratio of open space to the total volume of material. If the open space is permeable, then the permeability (K) represents the capacity of the porous medium to transmit a fluid (usually taken as length/time). Hydraulic conductivity is the capacity of a medium to transmit water. Permeability and hydraulic conductivity tend be used somewhat interchangeably in consulting work (since they refer to water transmission through natural materials). Groundwater moves under a hydraulic gradient (i) which represents the difference in height/length of incline. Hydraulic head is the difference in water level elevations measured in two wells under the hydraulic gradient. The relationship used to describe flow is Darcy's formula (see Figure 7):

$$Q = KAi$$

where $\quad$ Q = discharge
$\quad$ K = permeability
$\quad$ A = area
$\quad$ i = gradient

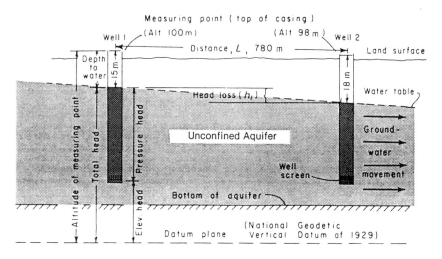

Figure 7.   Groundwater flow as observed in two wells showing head loss (h) over a distance (L) (Heath, 1982).

Transmissivity (T) is the capacity of an aquifer to transmit water through a unit thickness of the aquifer:

$$T = Kb$$

where b is the unit thickness of the aquifer. The storage coefficient is a dimensionless measure of water released from storage, where that volume = unit area × unit head change (see Figures 8–10). A more complete discussion of these relationships and aquifer tests used to measure transmissivity, the storage coefficient, and other parameters are presented in Chapter 8.

Aquifers are usually conceived as tabular bodies with distinct contacts or boundaries for ease in conceptual and mathematical presentation. At depth, given the properties of the individual aquifer, gravity and pressure forces (or heads) cause the flow. The potentiometric surface or pressure surface represents a plane of equal pressure relative to atmospheric in the aquifer as it flows from areas of recharge to areas of discharge. The flow will be toward areas of pressure head decrease in vertical or horizontal directions. Thus, initial estimates of flow direction made from topography may or may not be correct. In the case of an unconfined aquifer, the potentiometric surface corresponds with its groundwater occurrence. In the confined case, a confining stratum pressure surface and an imaginary surface may be

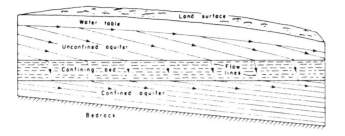

Movement of water through groundwater systems

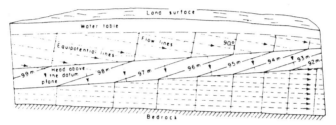

Figure 8. Groundwater flow from unconfined aquifer to confined aquifer through leaking aquitards represented by equipotential lines and flow lines (EPA, 1985).

made by observing water levels in wells that penetrate that aquifer. By connecting the points where the water rise is observed in wells, an imaginary plane is formed where the potentiometric surface would occur if not confined (see Figures 8 and 9).

Groundwater flow can be diagramed with flow nets. These flow nets are constructed by using equipotential lines, which connect points of equal head, and flow lines that represent idealized paths of water particles as they move in the aquifer. The flow nets may easily diagram any aquifer flow and the presence of lateral or upward gradients. The flow of groundwater may change direction and gradient given the recharge to, or withdrawal from, the aquifer. The gradient may change seasonally, and the assumption of permanent flow directions and gradients should not be made.

It is important to note that the aquifer is commonly imagined to be homogeneous and infinite in extent—this is rarely the case. These assumptions are made in order to quantify the aquifer and describe the flow in mathematics. Commonly, the geology of the aquifer is highly varied and will exert local influence by allowing or bounding

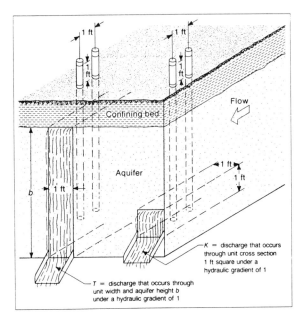

RELATIONSHIP OF AQUIFER HYDRAULIC CONDUCTIVITY (K) AND TRANSMISSIVITY
(T) THROUGH A UNIT AQUIFER (DRISCOLL, 1986).

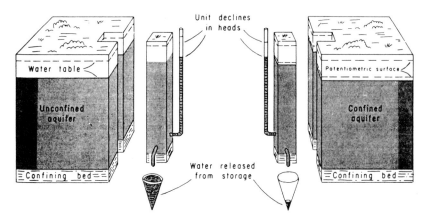

Figure 9. Water release shown by head declines in unconfined and confined
aquifers (Heath, 1982).

flow given the local situation. Data analysis must take the site-specific
geology into account when applying the appropriate mathematics to
analyze the groundwater flow.

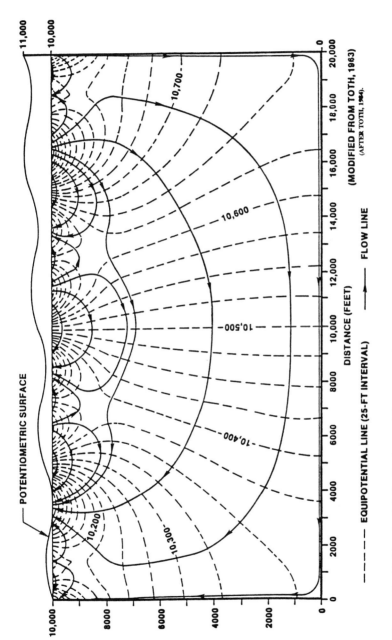

Figure 10. Fluid potential distribution and flow pattern in a theoretical flow system.

CONCEPTUAL APPROACH TO CONTAMINANT HYDROGEOLOGIC INVESTIGATIONS

A sensible geologic and hydrogeologic approach must be undertaken to determine extent of site contamination so remediation can occur. The site geology will form the physical framework through which the groundwater and contaminants flow. Thus, the geologic model should describe the hydrogeology, and the accuracy of the geologic study directly influences the accuracy of subsurface drilling, sampling, mapping, and the hydrogeologic study. Boreholes are expensive, and local access or legal restraints may limit placement of boreholes. Consequently, the information must be as complete and accurate as possible for each borehole drilled, given time, money, and the site conditions.

Consulting hydrogeologists may conduct contamination investigations on sites that are only hundreds or thousands of square feet (consider that an acre is 43,560 square feet). Thus, one is looking at very small areas in close detail, whereas classical groundwater investigations, such as U.S. Geological Survey (USGS) water supply studies, commonly encompass tens or hundreds of square miles. A contaminant investigation will be searching for shallow water saturated strata or zones that have probably not been previously used as water resources. Therefore, all the information will have to be generated from the site geological study. Very small features—including strata several inches or feet thick and potential contaminant pathways, such as animal burrows, plant root holes, or the results of human activity—can become highly significant. The "aquifers" of interest may only be a few feet thick, and those strata may occur as discontinuous beds or lenses with complex sedimentologic gradations. The field approach must be to collect the same information as one would for an areally large geologic investigation, but also the detailed and small-scale features.

As much information as possible should be reviewed prior to going into the field, including all the usual regional sources of geologic and hydrogeologic information in libraries or universities. The USGS is an excellent source of material for both geology and groundwater data contained in geochemical reports and the nationwide STORET computer data bank. EPA and individual states produce numerous technical guidance manuals and conduct research into specific contaminant-related studies, as do some universities. The quantity of research is growing; however, often information on contaminant transport and fate in geologic environments is limited.

This is only part of the local information needed for investigation. Previous site investigations and sources of contaminants should be located in relation to the study site. Other information sources are available and should be used since the site of interest may not be referenced in published material. Local planning offices at the county and city level archive soil engineering and geologic studies for land development, and much information on groundwater occurrence, soil type, exploratory boring logs, and well construction details are available as public information. Local water districts will have groundwater basin management data, such as producing well locations, groundwater pumping histories, and regional geology. With the growing hazardous waste management issues, the water district may also archive technical reports conducted specifically for contamination investigations. Local hazardous waste management agencies and fire departments at the city or county levels will have abundant information on sites that use or store hazardous materials, whether subsurface storage tanks are present or were removed, and if sampling and monitoring data were collected. This can be highly significant if the contamination has migrated from one site to another. Finally, the consultant may wish to discuss the site problem with experts in chemistry, regulations, law, and engineering. Contamination problems are complex and almost always require interdisciplinary approaches. Further, the regulations are more complex and definitive as to the types and completeness of information required. The more information one possesses, the more complete the problem definition and ability to formulate solutions.

AN EXAMPLE OF CHANGEABLE "AQUIFERS"

A spill site has been monitored with several shallow and deep wells. The initial aquifer interpretation appears to be thin and discontinuous sandy clay, which grades to a thick sandy aquifer. Aquifer behavior is puzzling since floating product (gasoline) and groundwater containing dissolved contaminants are periodically present in the shallow wells, which yield water very slowly. The deeper wells in the sandy portion of the aquifer are clean, yield water readily, and never contain floating product. When the potentiometric surface falls below the shallow wells, floating product does not enter the deeper wells as one would surmise. Also the dissolved product does not seem to migrate

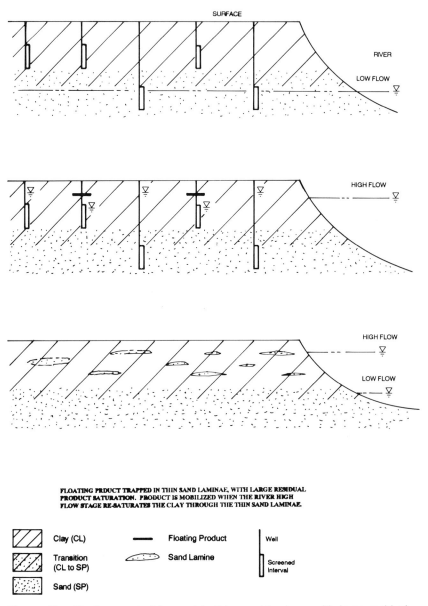

FLOATING PRODUCT TRAPPED IN THIN SAND LAMINAE, WITH LARGE RESIDUAL
PRODUCT SATURATION. PRODUCT IS MOBILIZED WHEN THE RIVER HIGH
FLOW STAGE RE-SATURATES THE CLAY THROUGH THE THIN SAND LAMINAE.

Clay (CL)   Floating Product   Well

Transition   Sand Lamine   Screened
(CL to SP)   Interval

Sand (SP)

Figure 11. Floating product trapped in thin sand laminae, with large residual
product saturation. Product is mobilized when the river high flow
stage resaturates the clay through the thin sand laminae.

in the direction suggested by the deeper wells, but rather the plume remains stationary.

The site is near a large river which rises and falls in response to precipitation and flood and irrigation dams release. As the surface in the river rises, the shallow wells yield water slowly with dissolved contaminants. When the river falls, the shallow wells display floating product and dissolved constituents, but the deeper wells are still clean. A cross section of the stratigraphy shows that the upper sandy clay "caps" the lower sand, which is always in communication with the river. When flow rises, it crosses the sandy clay–sand contact, and water appears in the shallower wells. The shallow wells are actually in a hydrogeologic unit that drains at low river flow. This allows contaminant to move slowly into the wells (and presumably vertically to the deeper sand). But the river flow is sufficient to prevent this less dense contaminant from penetrating across the contact, so the deeper wells are not yet affected. The groundwater movement of interest in the upper aquifer is really vertical rather than from up- to downgradient (see Figure 11).

Here the investigator needs to understand that the upper "aquifer" is only that when the river rise recharges, otherwise it is a vadose region above the "real" aquifer. Both low- and high-permeability sediments are involved. The observation of potentiometric surfaces is misleading if a layer cake interpretation is used. The gasoline movement is retarded since the rising and falling water and slow product movement in the sandy clay does not fall across the sandy contact. The alluvial geology controls the groundwater occurrence in the zone of interest (shallow wells), and the apparently anomalous groundwater occurrence is explained by both geology and proximity to the river. These factors helped keep spill somewhat confined, a fortunate occurrence for this case.

CHAPTER 2

Subsurface Exploration, Sampling, and Logging Techniques

INTRODUCTION

Prior to starting a field study program for a groundwater contamination project, different tasks should be addressed — ascertaining whether contamination is present, and meeting regulatory reporting requirements. Usually the responsible party will have been contacted by an agency or will do work in response to a spill, hazardous materials storage leak, or some other prompting order.

Initial work-scope decisions take into account the project goals, what type of investigation activity is undertaken, and budget and time constraints. Usually the consultant is working under some kind of client contractual arrangement in which a certain scope of work is defined and for which the client has agreed to pay. Thus, some type of bidding document has been devised for the specific work plan. This may be a voluminous work plan developed for a remedial investigation or a short three-page letter for soil sampling beneath a leaking tank. This work will usually have some kind of regulatory requirement, and the work should address this requirement fully. Logistical considerations in the work plan development are peripheral to actual hydrogeological study. In other words, the hydrogeologist must also know how to schedule, plan, budget, and select the appropriate equipment and contractors for the project. A review of some considerations for project work plans and startup are presented below.

SUBSURFACE EXPLORATION PROGRAM APPROACH

Geological Data Collection Goals

The subsurface program is the plan or scope of work that the consultant will use to collect information to map the site hydrogeology and to collect the physical information that the regulatory agencies will probably require. The regulatory agencies (federal, state, or local) will require information on the vertical and horizontal extent of "soil" and groundwater contamination at the site from the responsible party. To answer these needs, the number of exploratory boreholes, the number of samples to collect, and sample collection field methods must be decided prior to going into the field. Since the site-specific subsurface data collection will influence all the hydrogeologist's interpretations, the subsurface plan must be carefully thought out before any time, effort, or money is spent in the field.

The program should be designed to do the following:

1. establish the overall geological setting
2. allow definition of the site hydrogeology, such as which units are aquifers, which are aquitards, and so on
3. sample "soil" and groundwater at discrete intervals and points in both the unsaturated and saturated zones for both contaminant and geologic data

Thus, the hydrogeologist needs to estimate the number and position of exploratory borings and the number of monitoring wells, and to outline the soil- and water-sampling program (how many samples to analyze for which physical and chemical parameters). This should yield a clear understanding of the site hydrogeology and the relationship of geology to contaminant distribution within site vadose and saturated zones. It is very important to remember that this implies a completeness of information collected for the available project budget. The available money may limit the ability and volume of information, but should never compromise information quality and accuracy. The efficiency of data collection and coverage will be related to the professional experience and regulatory knowledge of the individual performing the investigation.

Consultants should use all possible information sources at their disposal to gain local site data for incorporation into the work plan, including the usual background geologic data sources such as USGS or state geologic and water resource reports on general geology and groundwater studies. Contamination investigation guidance documents that have been issued by state and federal agencies may help on investigation approach. However, often the site in question may not be near a previously investigated area—especially where new land development has occurred in the recent past. Thus, city and county planning office report depositories become vital for site-specific or nearby subsurface information from soil foundation engineering reports (soil boring logs, depth-to-water measurements, groundwater occurrence, and general geology). Since this information is almost always public, it can provide a wealth of data. When used in conjunction with city or county engineers, geologists, and water quality agency staff, an initial picture of site-specific conditions can be drawn. Additional related agencies, such as water districts or sewerage districts, may provide water quality and use information on a yearly basis.

Logistical and Drilling Contractor Selection Considerations

There are always logistical problems associated with field work, whether the site is downtown or on a remote hillside. The following kinds of problems may arise, typical of those with which the hydrogeologist must contend: available project budget, site access, permits, site clearance, and unforeseen financial problems.

The project can only be executed by spending the client's money. Although an emergency response might be done "without cost consideration," consultants usually operate on some kind of project budget. Consequently, each part of the project should be cost-estimated. Figure 1 shows a list of possible cost items that could be involved in an investigation. The consulting company is a business, and a company profit will be figured into the total budget. The client will review the budget and scope of work presented in the work proposal. The client must deal with the pollution problem but also wants the most yield for the money. Thus, the consultant must get the maximum information for each dollar spent. This also implies that the budget may constrain the information coverage available for data interpretation and problem definition.

Site access and permits needed for subsurface work are usually combined since they are needed for the field work to commence. Often well installation permits are needed for monitoring wells, just as they are for water supply wells. The permits may be required by city, county, or local district agencies and almost always must be approved before any work is started. A time delay may result from the permitting process and may need to be figured into the time schedule. The permit fee may be paid, and if necessary, field inspectors scheduled to observe the work. If the boreholes are to be placed in public easements, encroachment permits are needed for road or

```
                                              QUOTE NO:  _____
                                                    BY:  _____
                                      PROJECT LOCATION: _____
                            TIME x RATE $
PREFIELD
    Site walk              _____ hr x _____
    Research               _____ hr x _____
    Meeting                _____ hr x _____
    USA -- site clearance  _____ hr x _____
    Other (air photos, maps, etc.) _____
    Permits                _____ hr x _____        Subtotal _____

FIELD
    Drilling                    _____ hr x _____
        Prep to rig, cleaning   _____ hr x _____
        Steam-clean per hole    _____ hr x _____
        Casing    Blank         _____ ft x _____
                  Slot          _____ ft x _____
                  Grout         _____ ft x _____
        Finish Materials        _____
    Geologist log               _____ hr x _____
        Geotechnician           _____ hr x _____
        Sample materials        _____
        Travel -- Drill rig     _____ hr x _____
        Geologist               _____ hr x _____
        Vehicle                 _____ mi x _____
    Sampler                     _____ hr x _____
        Well development        _____ hr x _____
        Well sampling           _____ hr x _____
        Sample vehicle          _____ day x _____
        Travel                  _____ hr x _____
    V. and H. survey            _____          Subtotal _____
    Per diem                    _____ day x _____

CHEMISTRY ANALYSIS
    Soils        _____ x _____ per test(s) x 15%
    Waters       _____ x _____ per test(s) x 15% Subtotal _____

REPORTING -- Letter of Full Report
    Geologist        _____ hr x. _____
    Senior review    _____ hr x _____
    Typing           _____ hr x _____
    Reproduction     _____ ea x _____
    Drafting         _____ hr x _____
    Computer time    _____ hr x _____           Subtotal _____

MEETING
    Geologist        _____ hr x _____
    Engineer         _____ hr x _____
    Travel           _____ mi x _____            Subtotal _____
    Per diem
CONTINGENCY      Subtotals x 10%                       Subtotal _____
                                                       TOTAL    _____
```

Figure 1.   Example of a bid sheet. Project unit costs are totaled and presented for the bid proposal.

utility right-of-ways, and again a fee is charged. Permit fees are being used as a funding vehicle in some states for environmental compliance programs and may become widely used in the future.

Site clearance involves marking overhead and subsurface utilities and the need for safety plans on the site. Buried and aerial utility lines must be clearly marked and avoided to prevent costly damage and injury, especially to the drilling contractor. The site safety plan must be reviewed as to its appropriateness and completeness for the anticipated conditions. Safety plans are usually well-documented for any contamination investigation. A safety plan must be reviewed and approved prior to starting work regardless of the project type. A "standard" safety plan which meets basic state and federal requirements may form the basic safety plan, which is then revised as needs dictate. Worker protection (e.g., drillers, geologists, samplers) must be determined given the toxicological and exposure threat and requires the appropriate personal safety gear. These safety levels are outlined by Driscoll (1986) and are usually referred to as levels A (with full personal encapsulation and atmosphere required) through D (least restrictive). Obviously, the work should not commence until all safety considerations are met. A highly experienced and trained safety officer should assist the hydrogeologist consultant in preparing the safety plan.

Drilling contractor selection is probably the most important decision made to execute the subsurface exploration. The driller should be highly experienced with the chosen drilling method and construction of groundwater monitoring wells and preferably should have experience in the region of the investigation. Although drilling companies may be local to a region, the hydrogeologist must decide if these companies are capable of executing the scope of work. A statement of qualifications as to the company's ability should be solicited with the cost estimate. Also, the drilling contractor may be requested to bring all materials and equipment needed for the project, and this should be clearly negotiated prior to starting the field work to prevent delays and work stoppage. Subsurface work can be highly difficult and subsurface conditions are changeable, so an estimate of subsurface problems should be anticipated with which the driller must deal and a realistic time schedule for drilling, given the anticipated conditions, should be established. Driller selection is typically by bid, and the lowest bid is awarded the contract. However, the lowest bid may not be the "best," and although the financial factors are important, the most important factor is the drilling company's ability to get the job done on time and budget.

Lastly, the consultant and subcontractors may need to deal with insurance and bonding requirements. As the liability for subsurface and groundwater contamination work has risen in the past few years, so has the need for insurance and bonding. This may take numerous forms and is usually resolved in the negotiation and contractual stages of the work plan development with the client and contractors prior to going into the field. It may also involve the regulatory agencies and municipal bodies if they are letting the work. Although hydrogeologists may not be directly engaged in negotiating these problems, they should be aware of these conditions which may impose some restriction on the work, such as access to drilling locations, time schedules, and whether the work will be done by their firm.

SUBSURFACE DRILLING AND SAMPLING CONSIDERATIONS—INTRODUCTION

A number of drilling technologies are available for subsurface sampling and exploration. The techniques are designed for specific drilling conditions, and some methods are better than others, depending on the type of work envisioned. There is no universal drilling technique applicable for all subsurface conditions everywhere, so previous subsurface work experience becomes highly valuable. Driscoll (1986) presents an excellent review of drilling methods used in the water resource industry, and government guidance documents will review methods used in groundwater contamination work.

Sometimes the chosen drilling methods may be restricted due to site access, formation problems, and occasional regulatory requirements. However, ultimate selection of drilling equipment should always be made based upon the type of geology and sampling needs of the investigation. Subsurface drilling requires powerful, durable equipment, and availability of tools for the required hole diameter and for contamination (or any subsurface) investigation. The exploratory borehole is advanced so subsurface lithology logs and samples can be collected and monitoring wells installed. Depending on site-specific conditions, the drilling method should be chosen primarily on the ability to collect samples and maintain borehole stability, which is directly related to being able to execute the scope of work within the time and budget constraints to get the answers to contamination problems. Contrary to some popular opinions, the field work is not the easiest or simplest portion of the investigation, which can be done by "anybody." Field work should be reviewed and performed by hydrogeologists and project managers who are highly experienced in subsurface exploration techniques.

For this discussion, drilling techniques are broadly grouped into three of the most commonly used types for groundwater contamination work: augers, rotary, and cable tool drilling methods. It should be noted that these methods were initially developed for civil engineering, water resource, and mineral exploration uses. Since groundwater contamination investigation places a high premium upon cleanliness and decontamination, some accessory materials used on drill rigs should be carefully reviewed, such as drilling mud mixtures, rod greases, and cleaners and lubricants. If contaminants are inadvertently introduced into the borehole, the sampling data may be questioned.

DECONTAMINATION PROCEDURES

Decontamination of drilling equipment and sampling tools is vital to any groundwater contamination investigation. The purpose of decontamination is to clean the equipment of material that may cause contamination within or between boreholes. If these procedures are not used, or used incorrectly, the field data collected may become suspect and not defensible at a later date.

Decontamination may be overlooked in investigation planning. The level of decontamination will depend on the site problem and contaminant. A designated decontamination area should be selected, site walked, and approved prior to starting work. Peterson (personal communication, 1989) suggests the following in decontamination selection:

1. The decontamination area should be close to, yet not in, the study area. There should be all-weather accessibility and security if needed.
2. Utilities (power and clean water) should be available. Sometimes the only source of water may be a fire hydrant, or permission may be needed from (and billed by) the local water purveyor. The drilling operation should have a separate water source.
3. Fluids (both cleaning and drilling) will need to be disposed during and after the field work. The fluids may be disposed at approved and permitted sewer outfalls if available. Otherwise the final disposition may be according to hazardous waste disposal criteria depending on types and levels of contamination. This can become a significant cost and must be factored into the drilling budget.
4. All drilling and well construction materials must be cleaned. When-

ever there is a doubt about the cleanliness of the operation, activity should be stopped and decontamination procedures used.

5. As a general rule, there are two types of decontamination – hand washing and mechanical washing. Hand washing would be done to smaller sampling tools, jars, and reusable equipment. Approved cleaners (usually low sudsing phosphate, bicarbonate, carbonate, or hypochlorite) and clean water are used. Occasionally, a solvent rinse may be used before a deionized rinse. Mechanical washing usually involves a hot water or steam cleaner. This equipment uses high-speed jets of hot water to remove mud and drill cuttings, and volatilize contaminants. Usually this is the only method to quickly and efficiently clean the drill rig, kelly, rods, bits, and other tools. Cleaning in either case may be done on visquine plastic sheets and cleaned equipment or well construction materials should never be placed in contact with the ground or pavement.

SUBSURFACE EXPLORATION METHODS

Sampling and lithologic data are collected by advancing exploratory boreholes into the subsurface. The borehole may be advanced to any depth and into any type of formation or alluvium. Again, drilling method choice is dependent upon the type of formation and the anticipated conditions. Subsurface conditions are also changeable, and the driller and drilling method must be adaptable to compensate. Samples and lithologic logging are done while the borehole is advanced, and the hole is then backfilled with a seal, monitoring well, or other instrument as needed.

Auger Drilling

Auger drilling is usually the most common type of drilling method where sampling and drilling are fast and economical about 100 ft in alluvium and soft rock. Although auger rigs may have depth capabilities of 150–250 feet and deeper, when drilling to depths below 100 ft, the applicability of augers should be evaluated. Augers are versatile, very durable, and available in a wide range of diameters and fin and bit sizes. Flight augers and hollow-stem augers are usually used in initial investigations (see Figures 2–4 and Table 1). These types of augers and drilling methods are commonly used in the United States and are detailed by drilling rig manufacturers, government guidance documents, and Driscoll (1986). It is also prudent to contact the drilling equipment manufacturer as to the uses, capabilities, and limi-

Table 1. Advantages and Disadvantages of Auger, Rotary, and Cable Tool Drilling

Type	Advantages	Disadvantages
Auger	• Minimal damage to aquifer • No drilling fluids required • Auger flights act as temporary casing, stabilizing hole for well construction • Good technique for unconsolidated deposits • Continuous core can be collected by wire-line method	• Cannot be used in consolidated deposits • Limited to wells less than 150 ft in depth • May have to abandon holes if boulders are encountered
Rotary	• Quick and efficient method • Excellent for large and small diameter holes • No depth limitations • Can be used in consolidated and unconsolidated deposits • Continuous core can be collected by wire-line method	• Requires drilling fluids, which alter water chemistry • Results in a mud cake on the borehole wall, requiring additional well development, and potentially causing changes in chemistry • Loss of circulation can develop in fractured and high-permeability material • May have to abandon holes if boulders are encountered
Cable Tool	• No limitation on well depth • Limited amount of drilling fluid required • Can be used in both consolidated and unconsolidated deposits • Can be used in areas where lost circulation is a problem • Good lithologic control • Effective technique in boulder environments	• Limited rigs and experienced personnel available • Slow and inefficient • Difficult to collect core

Source: EPA (1989).

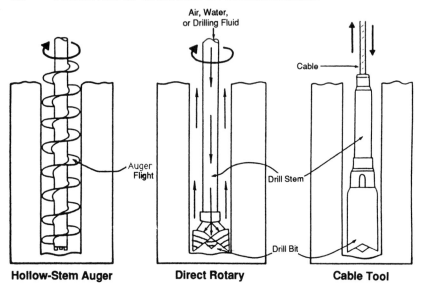

Hollow-Stem Auger   **Direct Rotary**   **Cable Tool**

Figure 2. Illustration and advantages and disadvantages of auger, rotary, and cable tool drilling. *Source:* EPA (1989).

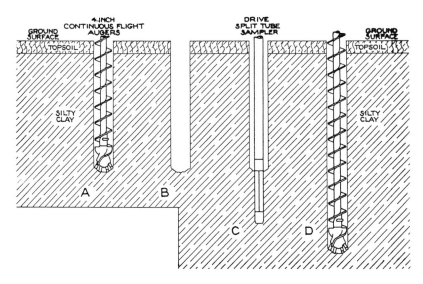

Figure 3. Continuous flight auger drilling. A. Advance auger to sampling interval; B. Remove flight augers; C. Advance split-spoon sampler; D. Advance auger to next sample interval. *Source:* University of Missouri, Rolla (1981).

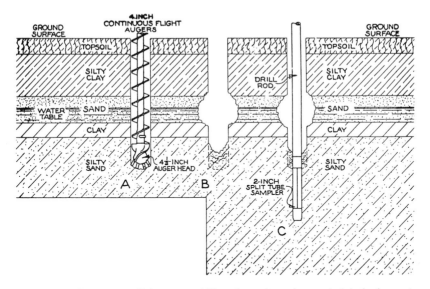

Figure 4. Continuous flight auger drilling through caving material. A. Auger to sample interval; B. Saturated sand stratum flows causing borehole to "bell"; C. Sampler must advance through sand "flow" slough to sample in-place silty sand. *Source:* University of Missouri, Rolla (1981).

tations of equipment. A summary of two types of auger drilling follows.

Flight Augers

Flight augers range from about 3 to 36+ in. in diameter, depending on the manufacturer, and are used to drill holes from 1 to 100 ft deep. Augers must be withdrawn to access sampling tools, and thus augers are typically used in soil engineering and sampling studies. They are very good for groundwater occurrence recognition because they use a "dry" drilling method. Borehole advancement problems are encountered when the formation caves or is unstable when the augers are withdrawn.

Hollow-Stem Augers

Hollow-stem augers are flight augers with a hollow core or stem, which varies from 3 to about 8 in., depending on the manufacturer. Again, these are used to drill holes up to 100 to 150 ft in most situations. Since these augers effectively "case" the borehole as the

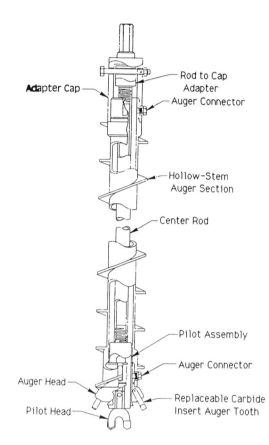

Figure 5. Components of a consolidated mining equipment hollow-stem auger. *Source:* University of Missouri, Rolla (1981).

hole is advanced, they help to prevent collapse or caving of the bore-hole. Sampling tools and monitoring well casing can be sent through the hollow stem by removing center plugs and rods (see Figures 5 and 6). This technique is excellent for water recognition and small-diameter well installation. It also can be used with soil-coring equipment for continuous soil sampling, and is also a dry drilling method. Borehole problems may occur when hollow-stem augers are advanced into unconsolidated, saturated sand formations, where flowing sand may enter the hollow stem and "lock" the augers in the ground. Auger removal can be very difficult since the borehole effectively has collapsed around the auger.

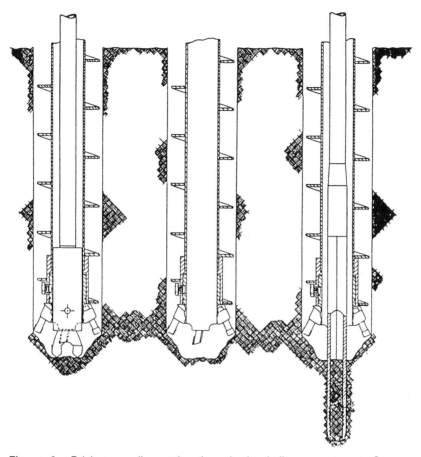

Figure 6. Driving a soil sampler through the hollow-stem auger. *Source:* University of Missouri, Rolla (1981).

Rotary Drilling

Rotary drilling drills a variable diameter (4-to 16-in.) hole from depths of tens of feet or thousands of feet. This technique uses a fluid circulated from a tank or pit down through the drilling rod and out the bottom of the bit. The rotary bit cuts through the sediment or rock, and the fluid is flushed to the surface between the drilling rod and borehole wall, moving the cuttings up and out of the hole. The cuttings then settle in the tank or pit, and the fluid is recirculated back into the hole (see Figures 7 and 8). This method is very versatile and can drill very deep holes. An added drilling advantage is that the

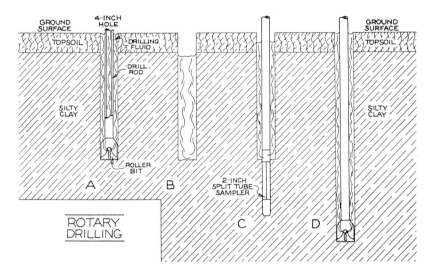

Figure 7.   Rotary drilling. A. Mud rotary drilling advance to sampler interval; B. Drilling mud holds borehole walls up; C. Split-spoon sampler advanced; D. New drill rod attached and borehole advanced. *Source:* University of Missouri, Rolla (1981).

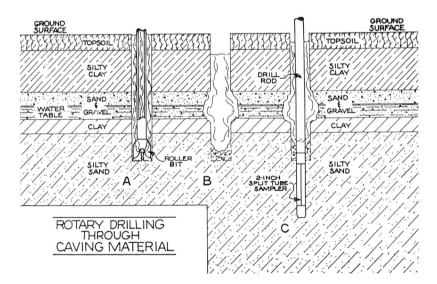

Figure 8.   Rotary drilling through caving material. A. Rotary drilling advances borehole to below water table; B. Drilling mud holes borehole walls open to minimize "flow" and caving; C. Split spoon advanced at desired sample depth. *Source:* University of Missouri, Rolla (1981).

drilling fluid, or mud, helps to keep the borehole open, and this is often the only method usable in caving or unstable formations. Samples can be collected at any interval by logging the cutting or chips of rock or sediment, or coring or drive sampling can be used. The technique can be difficult to use for water recognition since the borehole is already filled with fluid. This method does have the advantage of being used together with electric logs, utilizing the presence of borehole fluid and different logging tools for formation fluid identification. Variations on this method include air rotary, which uses air as the drilling fluid and high air velocities to lift the cuttings out of the hole, and reverse rotary, where cuttings and drilling fluid circulate in drilling rods.

Contrary to some popular opinions, mud rotary drilling is appropriate for groundwater contamination work. The mud will seal the borehole and water-bearing horizons effectively (Healy, 1989). If doubt exists, a drilling conductor casing can be advanced through an upper contaminated zone and mud rotary used thereafter. When experienced and qualified drillers are employed, this method is fast, economical, and highly adaptable to changeable subsurface conditions. Typically, when boreholes are to be advanced beyond 100 ft, mud rotary is the method of choice.

Cable Tool Drilling

The third technique is cable tool drilling, which is also the oldest method of drilling (Driscoll, 1986). This method drills a variable diameter (6- to 36-in.) hole to shallow or deep intervals by drilling about 4 to 5 ft using a hammering bit, then advancing casing to hold the formation, followed by another drilling interval. The method is essentially fluid-free, although some water is used to form a slurry to bail each interval's cuttings to the surface. Either a continuous chip log is recorded, or a discrete drive or core samples may be collected. This method can be used to advance a borehole through almost any formation and is especially useful in caving and collapsing formations that contain large boulders and cobbles. One disadvantage is that the method is often very slow, and casing must usually be advanced as the hole is drilled. Once the hole is complete to the desired depth, the casing may be removed, or left as the well or conductor casing.

Drilling Problems

Every time an exploratory borehole is advanced into the subsurface, different conditions could be encountered. The ability to advance the borehole depends upon the alluvium or formation, type of drilling equipment, and experience of the drillers. Although the work goes smoothly in most instances, changeable conditions can complicate, slow, or even halt drilling. The two factors that bear most on successful drilling are probably the type of equipment and experience of the operators. Problems will arise at some point when drilling on the job, and previous troubleshooting experience of the drill crew and geologist is invaluable. The drilling budget is usually the largest dollar amount, and the cost of labor and machines, given the difficulty of drilling, may require amending the field work. Examples of the more common problems follow.

Exploratory boreholes are usually advanced in increments (usually 5-, 10-, or 20-ft intervals) as the drilling stem is lengthened. The borehole condition may change as different material is encountered. Each material may behave differently when disturbed by the drill bit and hydrated by water, either from other strata or in drilling fluid. For example, clay may expand ("swell") when wetted, pinching and closing the hole. Sand may cave ("bell"), enlarging the borehole and vastly increasing the cuttings yield (see Figures 4 and 8). The longer the borehole is open, and drilling tools and equipment travel the borehole length, the more prevalent the erosion problem in the hole. After a prolonged period the borehole walls may become unstable and begin to collapse, and redrilling is required. A variation of this problem may include drilling through boulders or cobbles where the drill bit cannot cut or crush the particles. If boulder-sized or larger, the particle may prevent further borehole advancement, and a cable tool or percussion drilling technique may be required. In extreme cases, a casing must be driven through the bouldery stratum so the borehole can advance.

Another common problem is flowing sand. A flowing sand condition occurs where the sand is of relatively uniform diameter and saturated, and contains very little fine-grained material. The action of the rotating drilling bit causes the sand to liquefy and flow into the borehole. When this happens using hollow-stem augers, the sand may flow up into the augers, plugging and locking them into the formation. When rotary mud is used, the sand may flow and grab the drill stem, halting bit advance. When severe flowing sands occur, the usual solution is to use heavy or thick rotary mud whose weight stabilizes and holds the borehole open.

Finally, each borehole may have some limited amount of caving and drilling slough at the bottom. If severe caving conditions are a problem, the borehole—and ultimately well completion—may not be as deep as originally designed. A separate borehole drilled to the design depth may have to be advanced following the first which was used for sampling and lithology. A field judgment may have to be made by the geologist or hydrogeologist as to what acceptable amount of cave, borehole erosion, and so on is allowed.

SOIL-SAMPLING METHODS

A standard soil-sampling method has evolved over the last 30 to 40 years, the civil engineering soil-sampling technique (note that again "soil" means soil or unconsolidated sediment). Samplers used on both auger and rotary rigs use a driven sampling system, where the standard penetrometer sampler, sometimes called the *spoon*, is advanced by hammer blows ahead of the drill bit. The standard penetrometer is a 1.7-in. i.d., 18-in. long steel tube that collects soil for direct observation of texture and stratigraphy. The drive method uses a standard 140-lb hammer dropped over a 30-in. fall. The samplers are either steel tubes that can be split longitudinally to observe the collected sample, or simply steel, brass, or Teflon tubes. The number of blows to advance the sampler is related to indigenous strength characteristics of the soil sampled.

Variations on this technique include the California modified (1952) and California split-spoon samplers. The California modified sampler is a 2.0-in. i.d., 18-in. long steel-tube sampler, which contains tubes in which the sample is collected. The California split-spoon sampler is similar to the modified sampler except the inside diameter is 2.37 in. These samplers have been developed for use in soil engineering testing studies and are standard techniques that have been adopted for use in groundwater studies. Each sampler is driven in the same way. The sampler is lowered into the borehole, and a portion of the rod at the surface is marked in 6-in. increments. The hammer is then used to drive the sampler through an 18-in. length ahead of the bit, with the number of blows for each 6-in. increment noted. The actual sample is contained in the final 12-in. drive (the initial 6-in. drive commonly is through drill slough) so an "undisturbed" sample is retrieved. The blow count is recorded on the log, the sampler is dissembled, and the sample is lithologically logged (see Figures 2–4).

Additional sampling techniques are available to collect larger soil samples, depending on the type of information desired. The piston or Shelby tube sampler hydraulically pushes a 3-in. i.d. steel tube into the soil ahead of the bit. The tube is then retrieved and can be sent to the laboratory for geotechnical tests. This is a very common sampling technique for large soil engineering projects and is used increasingly for contamination work. The second method is "soil coring" and is similar to conventional hard-rock coring. The coring involves a diamond or hard-faced bit turned at high speed, or with hollow-stem drilling directly, and cores out the material directly into the core barrel. The barrel is then retrieved to the surface by a wire-line system, a second core barrel is sent down the hole, and the process repeated. The advantage of these methods is that "complete" sections of soil, sediment, or rock are recovered; the stratigraphy can be directly logged, and samples collected for specific interval analysis.

EXPLORATORY BOREHOLE LOGGING

Accurate borehole logging is essential for any subsurface investigation. The logging technique, as related to groundwater studies, is primarily a lithologic log of the soil, sediment, or rock. This log contains the primary observed subsurface information on soils, sediment, geology, groundwater occurrence, contamination occurrence, interval sampling, and aquifer and aquitard identification. If the lithologic logging is incorrect or incomplete, basic information relating to subsurface conditions could either be lost, be misinterpreted, or worst, go unseen.

Since some subsurface data collection techniques have grown out of civil engineering work, the logging style is a hybrid of both soil engineering and geologic logging. The geologic emphasis is needed, although the soil engineering "standard" is often used throughout the nation and is even codified in some laws and regulations. Since many investigations will deal with surficial soils and alluvial sediment deposits, the following discussion is primarily aimed at soil alluvium and sediment logging. Rock logging differs significantly from soil logging, and accepted logging techniques have been developed (for engineering, geology, and petroleum exploration uses).

The Unified Soil Classification System (USCS) is almost universally used for all soil engineering and engineering geologic work. The method is contained in ASTM D-2487 and 2488, and provides a textural classification and engineering performance of gravel, sand, silt, and clay (ASTM, 1988). These textures are similar to geologic texture classifications, and are based on standard U.S. sieve sizes, which differ only slightly from geologic. The USCS also classifies on the basis of soil engineering performance of moisture content, plasticity, and strength. Although these data may be useful for some aspects of the work, the primary emphasis is on groundwater and chemical contamination occurrence. Additional geologic information is required since the USCS does not necessarily take these purely geologic, hydrogeologic, and chemical features into account. For example, the presence of chemicals noted by odor, staining, or separate phase is by itself a primary reason for drilling the borehole.

Geologic data must supplement the USCS data for the log's usefulness to both engineers and geologists. The geologic information will provide direct insight into the hydrogeology, especially the porosity and permeability of the soil and sediment. This should include, at the minimum, bedding thickness, type and occurrence, sorting (opposite of grading), presence of biologic structures and secondary permeability features, and contact relationships between strata. These data are vital to observe and interpret grain-to-grain packing and the presence of fine material, which clogs porosity; indigenous structures that may allow cross-cutting openings for preferential contaminant migration; and the spatial relationship of each stratum to those that overlie and underlie the unit of interest. Obviously, the nature and relationship of aquifers and aquitards are the prime interest in these investigations, and without this information, the investigation is almost meaningless (see Figure 9).

A suggested logging procedure is presented that could be used to log boreholes for the groundwater contamination purpose. Since the borehole costs money to drill and the information collected from it is vital, it is the geologist's and hydrogeologist's responsibility to collect all the information that may be needed from that borehole. Thus, a system should be utilized that can collect the data to answer the questions at hand, the primary purpose of the consulting hydrogeologist.

The logging procedure would be accomplished as follows:

1. Drive the sampler (or collect samples) at the desired depth intervals. Record the blow counts or pull-down pressures.

Field location of boring:								Project No.:	12		Date:	08/10/89	Boring No:
								Client:	Oil Company				S-3
		(See Plate 2)						Location:	1702		Street		
								City:	, California				Sheet 1
								Logged by: S. Carter			Driller: Bayland		of 2
								Casing installation data:					
Drilling method: Hollow-Stem Auger													
Hole diameter: 8-Inches								Top of Box Elevation: 107.05			Datum: MSL		

PID (ppm)	Blows/ft. or Pressure (psi)	Type of Sample	Sample Number	Depth (ft.)	Sample	Well Detail	Soil Group Symbol (USCS)	Description
								Water Level 12.60
								Time
								Date 08/11/89
								PAVEMENT SECTION - 6 Inches
			1					CLAY (CH) - very dark gray (2.5Y 3/0), stiff, moist;
				2				10-15% coarse sand.
				3				
				4				Note: Switch to GSA Rock - Color Chart at 4.0 feet.
	6/3 "							GREAT VALLEY SEQUENCE - Unnamed Formation-
	32			5				SILTSTONE/MUDSTONE - grayish olive green (5GY 3/2),
191	43	S&H	S3-5.0					well indurated, massive with thin (<1/4") interbeds of
				6				clay; strong chemical odor and discoloration.
				7				
				8				
				9				
	14							color change to dark yellow brown (10YR 4/2) at 9.0 feet,
	34			10				no clay interbeds.
3	40	S&H	S3-10.0					
				11				
				12				
				13				
				14				
	6							SILTSTONE/FINE GRAIN SANDSTONE - weakly
	30			15				indurated, dary gray (N3), closely fractured
0	37/2"	S&H	S3-15.0					MUDSTONE - moderately yellow brown (10YR 5/4),
				16				friable, closely fractured with calcite on fracture surfaces
								at 14.5 feet.
				17				SILTSTONE - weakly indurated, dark gray (N3), closely
				18				fractured; no chemical odor.
				19				
	11							
0	50/3"	S&H	S3-19.5	20				SILTSTONE - strong, dry, closely fractured, (10YR 4/4),

Remarks:

Log of Boring   BORING NO

Figure 9. Example of an exploratory boring log, with both soil and rock logging descriptions.

2.   Remove the sample from the sampler in correct orientation. Package and seal any samples retained for chemical analysis to prevent volatility. The method of packaging for chemical analysis should be appropriate for the suspected contaminant.

3.   Lithologically log the soil or sediment. A suggestion for completeness is to always log in the same order. Always logging in

the same order makes collecting the minimum information almost automatic but should not be used as a substitute for collecting additional data as needed on a site-by-site basis. The data are collected in this order: texture (USCS), color (Munsell), percentage of constituents, plasticity of silt or clay, presence of bedding, presence of biogenic structures, presence of chemical vapors and staining, consistency and/or density (from ASTM method), estimated relative moisture content (i.e., damp, moist, or saturated), other features.

One must remember that this logging technique should not be used for rock logging. The USCS classification was developed for classification of soil geotechnic properties of unconsolidated deposits. Consequently, if one is logging rock, the log should collect information needed for consolidated or indurated formations. The formation name and the contact relations between the overlying sediment and each underlying formation or stratigraphic unit should appear on the log. Petrographic, fracture spacing, joints, shears, rock quality, presence of fossils, contaminants, and other information is typically included. Accepted rock logging classifications have been developed by engineering geologic, mining, and petroleum companies. These can be adapted to the site-specific study.

Logging aids are available which the geologist or hydrogeologist should use to increase accuracy, including a hand lens for close examination, color chart references (such as the Munsell and G.S.A. color charts), grain size and percentage aids, see-through ruler, and a protective notebook. An aluminum notebook is recommended since often maps, permits, safety information, and other documents are carried in the field and should be protected.

Detection of Groundwater in the Exploratory Borehole

The detection of groundwater is not always clear, and typically the driller may advance into water before a collected soil sample marks the occurrence. The following suggestions will aid in logging first groundwater occurrence, which is one of the most significant pieces of data desired. They have been complied from the combined experiences of drillers and geologists:

1. Always collect a soil sample if water is thought to occur at a soil interval of interest. This should be done so the hydrogeologist sees

the occurrence in the soil, and extra samples should be driven to be sure if needed.

2. Changes in drilling resistance or penetration rates can be useful. The rate of penetration may decrease if moist conditions are encountered. Rates of drilling resistance are almost always logged and should always be noted.

3. Increasing changes in observed soil relative moisture contents with depth may imply approach of an aquifer contact.

4. Once drilling in a saturated unit, a sample should be taken to see if the drill bit has advanced into an underlying "dry" unit or aquitard. This is extremely important to prevent cross-connecting two aquifers, especially if the overlying aquifer is contaminated.

5. Geologists should talk to the drillers and discuss what they want to do and anticipate during the course of the drilling. The drillers are usually widely experienced, and although they may not have degrees in hydrogeology, they do have abundant subsurface experience and a feel for drilling operations. This can be very important if the driller has worked in the area and the geologist has not. The geologist must log the borehole, but the driller can aid in subsurface information collection.

6. Once water is encountered, water entry may be slow, and the drilling may need to be stopped to ascertain if water is, in fact, present. This is important when working in low-permeability situations.

7. Measurement of depth to groundwater should be taken when it is first encountered and compared to later depth measurements to observe potentiometric rise.

8. Packer tests may be performed at selected intervals. The packer test seals off an interval of the formation to allow pumping out, or pumping in, water tests through the drill rods. These are especially useful in consolidated and fractured formations.

Subsurface information collection is the heart of any investigation since all subsequent interpretations will be based upon the lithology log compiled by the hydrogeologist. Consequently, the interpretations will only be as accurate as the data collected. Hydrogeologists and geologists must be highly experienced in borehole logging and recognition of significant geologic features. Logging numerous boreholes increases the individual's ability; each borehole is different, and new insight is gained from logging experience and assists field personnel when supervised from the office. As in other endeavors, there is no substitute for experience in the field.

GEOPHYSICAL LOGGING

Geophysical logging techniques have been used for many years in mining and oil exploration work, and are finding use in groundwater investigations. Conventional electric logging has been performed for years in water resource investigations. Usually geophysical logging is used to aid in collecting additional lithologic data and filling in gaps that the visual log may have missed, or in areas where samples cannot be retrieved. Thus, electric logs make sense when drilling is very complex, at great depths, in unstable formations. Commonly, resistivity, spontaneous potential, and gamma logging are used to resolve lithology. A review of geophysical logging is beyond the scope of this book; however, a brief discussion is included since the hydrogeologist may need to use the technique.

Typically, firms specializing in electric logging are employed to provide the service and at times for log interpretation. Several techniques are commonly employed for lithology identification, and tools may provide additional information on borehole conditions, including borehole "decay" or caving, deflection from the vertical to calculate true borehole bottom location, and consistency of borehole width with depth. Another use is that sometimes borehole data are needed from previous drilling operations and borehole logs are not available. One tool, the gamma log, may be used to provide a general log of occurrence of sand and clay beds, including through completed wells. Thus, a rough log can be collected from a well installation point without going to the additional cost of drilling another hole adjacent to that well (see Figure 10).

EXAMPLE OF INTEGRATING SAMPLING AND DRILLING

Suppose you are retained to drill an exploratory boring 650 ft deep and install the smallest-diameter, fully penetrating monitoring well that you consider feasible given the available budget resources. The site geology consists of alluvium to depths of about 100 ft, underlaid by a weakly consolidated formation. You have to collect samples of the alluvium for chemical analysis, but the contamination does not appear to have penetrated into the formation, based on previous information. You want to get a good idea of the formation stratigraphy and retain samples for permeability analysis. Also, an aquitard unit occurs at 450–600 ft, and a pumping aquifer underlies

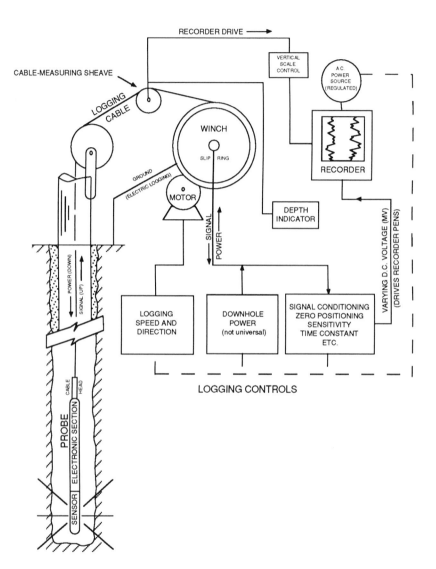

Figure 10. Schematic diagram of geophysical logging. *Source:* Welenco (1985).

the aquitard. Finally, the budget for the project is not large, and although accuracy in logging is primary, the project must be done as fast as possible. How do you proceed?

It is probably wise to approach this in two phases: boring hollow-stem augers to the contact of the alluvium and formation, then switching to a mud rotary method of drilling. Hollow-stem augers will usually advance to 100 ft, and since you need to go another 550 ft, mud rotary will be the fastest with sample coring capability. In this way, you have the nonfluid soil-sampling ability needed to collect samples for chemical analysis and the power and depth of penetration rotary affords. The weakly consolidated formation may have a tendency to cave or have severe borehole wall erosion, so fluid drilling should give you the best method to advance to the desired depth and complete the well. Also, since you are drilling a deep hole, you budget electric logs so sand and clay beds can be located and to supplement the visual "chip" log (see Figure 11).

The hollow-stem auger borehole is advanced, and soil samples are collected at the desired intervals, including at the alluvium and formation contact. This should give you the chemistry profile the client desires. Now, this auger hole provides the pilot borehole for a second, slightly larger auger borehole into which a steel conductor casing will be installed. Remember, the alluvium is at least slightly contaminated, and if mud rotary is initiated immediately, some contaminants may be circulated deeper. Hence, the conductor is lowered, pressed into the formation, and cemented into place. The grout is allowed to set before rotary drilling starts so the seal is not disturbed.

Now the mud rotary drilling may start. The drilling will now yield a continuous lithology chip log so formation conditions can be viewed at all times. Although the continuous chip log shows lithology, it is wise to core episodically, so given the time and money (since coring is expensive) we will core 20 foot intervals every 100 ft as we approach the aquitard. This provides you with samples that will be used for laboratory permeability testing, as well as lithology checks on the electric logs. Since you are aware that an aquitard is present at 450 ft, you should core an interval from 440 to 460 ft to verify the upper contact and to allow you to view the contact relationships. The nature of the aquitard is critical to providing the aquifer protection, so additional core samples will be collected from 510 to 530 ft and from 580 to 620 ft to view the internal stratigraphy and the aquifer-aquitard contact. The design depth of the well is to 650 ft, so the borehole is advanced to that depth. The aquifer is much sandier and fine-grained than the literature has implied in this area, so a complete log is needed. Hence, while the borehole is open, you had the foresight to schedule an electric logger to log the hole, so now you will have two complete logs. The logger must perform the task quickly since the circulating mud will erode the aquifer section of the borehole. Well

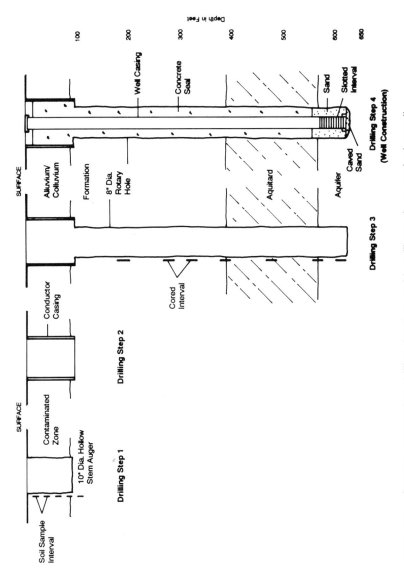

Figure 11.   Possible approach to drilling, logging, and installing a deep monitoring well.

construction must immediately follow the electric logging. When the well casing is lowered, you discover that about 10 ft has caved at the base of the hole (now 640 ft). The driller has informed you that while he may be able to clear the lower 10 ft, the borehole may cave to 600 ft (since the aquifer is so sandy), forcing redrilling that interval under flowing sand conditions. Given the time and cost, and now borehole instability, the decision is to accept the loss of 10 ft rather than lose the borehole, and so the well is constructed at that point.

Have you performed the tasks assigned and contractually agreed to by you? The alluvium has been logged and chemical analytical samples were collected, the drilling proceeded only after the contaminated zone was sealed off, permeability and lithologic samples were collected, electric logs fill other lithologic information gaps, and the well was constructed. The well was 10 ft shorter than desired, but given the possibility of losing the entire aquifer portion of the hole (and all the time and money spent therein), it seems a good compromise. By using your experience and judgment, the work scope goals have been met, yielding the information required for the study.

CHAPTER 3

Groundwater Monitoring Well Installation

INTRODUCTION

Groundwater monitoring wells are data collection points installed during the investigation. The wells will provide site groundwater potentiometric and quality information for investigation and for future and closure monitoring. Monitoring wells will be used for some indefinite time period (usually years), so construction materials selection and installation criteria are very important. Consequently, the effort expended in installing the system should be well thought out, and the system installed carefully. Improperly designed or located wells may lead to numerous problems, such as improper plume location, cross-connecting aquifers, and questionable flow and quality information—costing time and money in terms of data lost or questionable data for which new wells may be required.

Monitoring wells should be located and constructed to collect representative samples of groundwater quality and provide reliable potentiometric data. The design of the well should always reflect the site-specific hydrogeologic environment. Monitoring well design should be initially be done in the office, given the known or surmised site-specific geologic conditions. Well installation permits may require the consultant to provide a schematic or conceptual design of the proposed well prior to installation. At times regulatory agencies may review, modify, or even specify monitoring well design, which the consultant may need to use as a requirement for obtaining an operating permit or part of the investigation work plan. This can predictably create problems when fine points of design, or geologic realities, make design changes necessary beyond arbitrary dictums or generalized cases provided in government guidance documents. Finally, the type of contaminants and natural geochemical conditions will have a bearing on materials used and positioning of well screen or open casing sections.

The finalized design will depend upon previous experience, regulatory guidance, geologic conditions, and professional judgment (Driscoll, 1986). Since all drilling and well completion operations are complex and detailed, the hydrogeologist should be onsite to witness all phases of drilling and construction completion. Cross-contamination of aquifers or construction that would enhance contaminant movement must always be avoided and, without considerable diligence, may happen to even the experienced hydrogeologist (EPA, 1986a; Hackett, 1987). Field experience is required and is vital for success in this work. These projects cannot be properly managed from the office alone and require experienced field personnel directly observing the work.

ARTIFICIAL FILTER (SAND) PACK SELECTION

If native aquifer materials cannot be used as a filter pack, an artificial sand pack is usually required. The use of an artificial sand pack is particularly useful in environments with uniform fine-grained sand, poor cementation of sand, stratification within aquifers, or where a long screen interval is required. The sand pack also stabilizes the formation (preventing caving) while allowing water entry into the well.

Design criteria for sand packs involve review of aquifer sieve and proposed well screen information, but not necessarily a set of "standard" design steps. Several methods, developed for water production wells, may be used, and the reader is referred to those publications (U.S. Department of Interior, 1981; Driscoll, 1986). Artificial sand pack design usually involves collecting samples of aquifer material and performing a grain size sieve analysis to ascertain the grain size distribution in the aquifer. Once the aquifer grain size distribution is known, then the effective size of the aquifer material materials can be determined. Usually the sand pack selected must be slightly larger to retain the formation while allowing for a screen slot size sufficient to allow water entry. Finally, the sand pack should be clean, typically using 90% quartz to eliminate reactive minerals and composed of rounded grains, which increase yield to the well. Aggregate and sand suppliers can then prepare the sand to the design specifications.

Usually, at least one aquifer sample sieve analysis should be done, and the results used for monitoring well sand pack. The sand pack design will involve a judgment based on the sieve data and uniformity coefficient to select a filter pack grain size. The uniformity coefficient is the ratio of the sieve size that allows 60% (by weight) of the material to pass to the sieve size that allows 10% of the material to pass. The sieve data are interpreted using the range of the aquifer gradation uniformity coefficient. Driscoll (1986) recommends that the filter pack for monitoring wells should have a uniformity coefficient of 1 to 3, (multiplying the 70% retained of the finest formation sample by 3 or 4). Other guidance indicates that, depending on the range of the uniformity coefficient, the sand pack is selected to retain 50% of the formation by multiplying the 50% size by a factor (U.S. Department of Interior, 1981). The sand pack selected should then have a range of grain size that is somewhat larger than the aquifer material and slightly larger than the well screens. Thus, following well development, sand pack should grade from coarser near the screened interval to finer into the formation.

A dilemma exists when the aquifer strata contains a large fine-grain constituent (silt and clay), where the design criteria indicate a screen size and sand pack so fine that it may inhibit water entry into the well. Consequently, the design criteria must use an analysis of the sieve data compared to the aquifer material, and be tempered by professional judgment and previous well design experience. Once aquifer conditions and site geology are known, then a similar design might be used for additional monitoring wells in the site vicinity.

WELL CASING AND SCREEN MATERIALS SELECTION

Materials selection for wells is very important in design. Numerous articles and tests on the different types of materials have been done and other work is ongoing. EPA guidance usually ranks available well materials, in order from best to worst, as Teflon, stainless steel, polyvinylchloride (PVC) type 1, low carbon steel, galvanized steel, and carbon steel. For this discussion, three materials are considered — steel, PVC, and Teflon — since these are currently the most used according to EPA and a poll by McCray (1986). Interestingly, McCray's poll found that 93.3% of the respondents indicated that PVC was the material of choice. These three materials are used in monitoring at large or small sites, each having its advantages and disadvantages in certain geologic and contaminant environments. Beyond construction material and service life considerations, the potential to release or adsorb contaminants is the primary design criterion. For example, steel casing used in extremely saline or cathodic environments may

Table 1.   Comparison of Well Casing and Screen Materials (1986)

Type	Advantages	Disadvantages
PVC (Polyvinylchloride)	• Lightweight • Excellent chemical resistance to weak alkalies, alcohols, aliphatic hydrocarbons, and oils • Good chemical resistance to strong mineral acids, concentrated oxidizing acids, and strong alkalies • Readily available • Low priced compared to stainless steel and Teflon	• Weaker, less rigid, and more temperature-sensitive than metallic materials • May adsorb some constituents from groundwater • May react with and leach some constituents from groundwater • Poor chemical resistance to ketones, esters, and aromatic hydrocarbons
Polypropylene	• Lightweight • Excellent chemical resistance to mineral acids • Good-to-excellent chemical resistance to alkalies, alcohols, ketones, and esters • Good chemical resistance to oils • Fair chemical resistance to concentrated oxidizing acids, aliphatic hydrocarbons, and aromatic hydrocarbons • Low priced compared to stainless steel and Teflon	• Weaker, less rigid, and more temperature-sensitive than metallic materials • May react with and leach some constituents into groundwater • Poor machinability—it cannot be slotted because it melts rather than cuts
Teflon	• Lightweight • High impact strength • Outstanding resistance to chemical attack; insoluble in all organics except a few exotic fluorinated solvents	• Tensile strength and wear resistance low compared to other engineering plastics • Expensive relative to other plastics and stainless steel

Table 1. Continued

Type	Advantages	Disadvantages
Kynar	• Greater strength and water resistance than Teflon • Resistant to most chemicals and solvents • Lower priced than Teflon	• Not readily available • Poor chemical resistance to ketones, acetone
Mild steel	• Strong, rigid; temperature-sensitivity not a problem • Readily available • Low priced relative to stainless steel and Teflon	• Heavier than plastics • May react with and leach some constituents into groundwater • Not as chemically resistant as stainless steel
Stainless steel	• High strength at a great range of temperatures • Excellent resistance to corrosion and oxidation • Readily available • Moderate price for casing	• Heavier than plastics • May corrode and leach some chromium in highly acidic waters • May act as a catalyst in some organic reactions • Screens are higher priced than plastic screens

Source: Driscoll (1986).

release alloy metals. PVC may soften and decay in environments containing high concentrations of certain organic solvents. Teflon may yield metals in certain acid environments. Consequently, the casing choice is that which best adapts to the problems of water quality, geology, and budget (Table 1).

Leaching tests by Sykes et al. (1986) and Barcelona and Helfrich (1986) show that in material exposure experiments, significant differences did not exist for these materials in terms of leaching and adsorption. Thus, the casing selection for some sites, with other factors being equal, is often a financial consideration. Since the typical cost of steel may be four times PVC, and Teflon up to ten times PVC, the project budget factors into the construction criteria. However, the cost basis should be made given the estimated chemical exposure and service life.

LENGTH OF WELL SCREENS

The length of the well screen is dependent upon the aquifer thickness and contaminant type for site monitoring. Screen length is dependent upon formation and contaminant conditions. The screen should allow the well to yield water and optimize well efficiency, sample the contaminants of interest, allow for seasonal groundwater fluctuations, and enable hydraulic head data to be collected. For example, monitoring wells installed to monitor immiscible contaminants should be screened somewhat above "static" water levels to allow separate phase product entry into the well. Wells may be discretely screened to observe stratified or sinking contaminants in different aquifer intervals.

Ideally, a well screen should penetrate the aquifer for the following reasons (Peterson, in Behnke, 1990):

1. to increase the well yield and efficiency of development
2. to allow for observation and capture of contaminants
3. to be useful during seasonal groundwater variations
4. to be effectively and accurately utilized during aquifer testing for characterizing aquifer parameters

Usually, fully penetrating well screens are used in initial site studies. Screen lengths are later modified in additional wells, given the delineation of the aquifer and contaminant. For example, well screens may be designed to be 10 ft long when looking for stratified contaminant flow, or lengthened above the capillary fringe for free product observation. However, no selection formula exists, and screen length should always be tailored to the need at hand, providing aquifer cross-connection and other problems are prevented.

ANNULAR SEALS

The well annulus (that region between the well casing and borehole wall) must be sealed to isolate the screened interval and prevent contaminants from entering the well directly down the exploratory borehole. Since the seal is required to be "impermeable," usually a bentonite seal, a cement grout seal, or bentonite-grout mixture seal is used. The bentonite seal is placed directly above the sand pack and initially seals the annulus and will prevent grout invasion into the sand pack. A trémie line, or pipe, is used to deliver the cement to the area to be sealed. The cement grout is pumped by trémie line atop the bentonite and continuously fed from bottom to top into the annulus until grout appears at the surface. If a bentonite-grout mixture is used, usually between 5 to 7% bentonite is mixed into the grout. Grout mixture design may vary with the project, as directed by state or agency jurisdiction, and subsurface conditions (such as reactive minerals). However, either premixed bags or aggregate plants can easily prepare the grout mix specification. Grouts will settle following placement and are usually topped off when the surface wellhead access is constructed. Most grouts will set in 12 to 24 hr (Table 2).

SURFACE COMPLETION

The well surface completion provides ground surface access and security at the wellhead. Typically, most monitoring wells are completed below ground surface and are accessed through a traffic-rated concrete box. If wells are accessed aboveground, then marking pipes and traffic barriers may be needed to mark and protect the wellhead. Wellhead casings rise through surface casing or "stovepipe collars," which are locked and help to prevent unauthorized entry and vandalism. A metal tag containing well information, such as total depth and date of installation, may be wired to the surface casing. Final completions may be further tailored to the needs of the site, client, or regulatory agency requests.

SUMMARY

Monitoring well design must take into account site geology at the well location, type and position of contaminant, input from regulatory agencies, and past well installation experience. Contingencies for well design modification in the field should be anticipated if subsurface conditions are significantly different from those anticipated. The screen length of each well should be site-specifically determined given aquifer conditions and contaminant type. The well sand pack should be designed for the aquifer texture and may be used in all other well designs if significant aquifer textural changes are absent. An impermeable seal is placed atop the sand pack; a continuous grout seal is finished at the surface with a locking access box. The final well design and care of construction take these considerations into account, so the monitoring well is usable for its anticipated service life.

Table 2.   Comparison of Well Grouting Materials

Type	Advantages	Disadvantages
Bentonite	• Readily available • Inexpensive	• May produce chemical interference with water quality analysis
		• May not provide a complete seal because —There is a limit (14% to the amount of solids that can be pumped in a slurry. Thus, there are few solids in the seal; should wait for liquid to bleed off so solids will settle
		—During installation, bentonite pellets may hydrate before reaching proper depth, thereby sticking to formation or casing and causing bridging
		—Cannot determine how effectively material has been placed
		—Cannot assure complete bond to casing
Cement	• Readily available • Inexpensive • Can use sand and/or gravel filter • Possible to determine how well the cement has been placed by temperature logs or acoustic bond logs	• May cause chemical interferences with water quality analysis • Requires mixer, pump, and trèmie line; generally more cleanup than with bentonite • Shrinks when it sets; complete bond to formation and casing not assured

Source: Driscoll (1986).

MONITORING WELL CONSTRUCTION: AN EXAMPLE OF INSTALLATION

The following example is given as a typical shallow (40–50 ft) monitoring well installation, so that the basic construction sequence is illustrated. We will assume that the borehole has been advanced to the desired depth, lithologic logging has determined where the screened interval will be placed, and the contaminant type and geology were factored into material selection. Finally, we assume that the hydrogeologist has finalized the design, and the proper material is cut to specifications. The field hydrogeologist must accept the ultimate responsibility for the final design and correctness of construction.

The site hydrogeologist must observe the casing prior to installation to look for potential contaminants and clean the casing if needed. Glues and epoxies are never used in construction since they can yield contaminants similar to those for which you might be searching. Since casing is commonly manufactured in 5-, 10-, and 20-ft lengths, the casing is threaded together and lowered into the hole in those increments. The casing would be lowered through either hollow-stem auger or thinned rotary mud, depending on the drilling. An end cap or plug is fitted or screwed to the bottom of the screened interval or silt sump to prevent pumping of aquifer material beneath the well. Tension is kept on the casing string to prevent kinking, and well centralizers may be used to keep the casing centered in the borehole. Once all the casing has been lowered into the borehole, then the annular space between the casing and borehole is filled with the sand pack and seal.

The sand pack installation must be done carefully to avoid sand bridging and ensure the sand is properly delivered about the screen. The sand is often sent down by a trémie pipe to the required depth and really should not be allowed to air fall except in shallow (less than 50–100 ft) completions. The sand pack will fill the annulus, which should be a minimum of 2 to 4 in. between casing and hole wall. Once emplaced about the screened interval, the sand is typically filled to a point about 2 ft above the screen. This ensures the screen is completely covered and creates a spacer above the screen for the impermeable seal above it.

The annular seal will isolate the well screen into the aquifer and preclude communication with other aquifers. Typically, a bentonite seal (about 2 ft thick) is placed above the sand pack prior to grouting the annulus. This is very important since grout invasion of the sand pack may occur, rendering the well useless. Using bentonite at this point also forms a seal in the annulus and isolates the aquifer immediately. Usually a bentonite pellet product is used, which sinks into position, and the pellets expand upon hydration to effect the seal. If bentonite cannot be used due to extreme depth, then a 5- to 10-ft sand pack spacer should be placed to absorb the grout invasion that will occur. This allows flexibility in construction at times when well construction is very difficult.

Once the bentonite is placed, a grout seal is trémied into position in the annulus and completes the seal. The grout is pumped downhole until the space is filled to the ground surface. Grout setting may adversely affect nonsteel casing, and the grout may need to be placed in intervals or lifts to provide for casing strength considerations. Again, the care of construction should be the same as that of any other part of the geologic and hydrogeologic study. The site hydrogeologist should observe the entire construction sequence and note additional need of material (such as a need for extra sand or grout if the borehole has become unstable, or difficulties in lowering casing, which may smear and seal screen openings).

Finally, the surface completion must be done to protect the well surface casing from flooding and vandalism. A secure cap with locking device should cover the wellhead, which, in turn, is contained in a traffic-rated vault box. If the wellhead is aboveground and in an area used by traffic or in an open field, a concrete pad surrounded by guard posts may be used. The well access should be graded slightly so positive drainage is provided away from the well. Additionally, information regarding well completion (well number, depth, date of installation) should be written on a metal tag (see Figures 1 and 2).

MONITORING WELL DEVELOPMENT

All wells must be developed following installation. Well development accomplishes the following:

1. It clears suspended particles from the water column.
2. It removes mud cake and smeared material on the borehole walls from drilling—especially important in rotary drilling.
3. It grades the sand pack into more complete contact with the aquifer.
4. It provides the hydraulic connection between the casing and aquifer.

The development process usually is done by one of two methods: either surging and bailing, or jetting. The surge-and-bail method involves running a bailer or surge block down into the well and retrieving a volume of water and suspended sediment and bringing it to the surface. Aquifer water will then move into the evacuated well, flushing out additional sediment. The development process is repeated until the well water becomes clean or meets a specified water clarity criterion. The surging action moves water through the well openings,

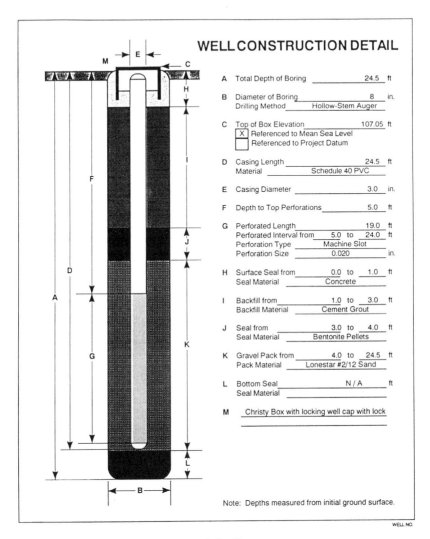

WELL CONSTRUCTION DETAIL

A   Total Depth of Boring _____ 24.5 ft

B   Diameter of Boring _____ 8 in.
    Drilling Method _____ Hollow-Stem Auger

C   Top of Box Elevation _____ 107.05 ft
    [X] Referenced to Mean Sea Level
    [ ] Referenced to Project Datum

D   Casing Length _____ 24.5 ft
    Material _____ Schedule 40 PVC

E   Casing Diameter _____ 3.0 in.

F   Depth to Top Perforations _____ 5.0 ft

G   Perforated Length _____ 19.0 ft
    Perforated Interval from _5.0_ to _24.0_ ft
    Perforation Type _____ Machine Slot
    Perforation Size _____ 0.020 in.

H   Surface Seal from _____ 0.0 to _1.0_ ft
    Seal Material _____ Concrete

I   Backfill from _____ 1.0 to _3.0_ ft
    Backfill Material _____ Cement Grout

J   Seal from _____ 3.0 to _4.0_ ft
    Seal Material _____ Bentonite Pellets

K   Gravel Pack from _____ 4.0 to _24.5_ ft
    Pack Material _____ Lonestar #2/12 Sand

L   Bottom Seal _____ N / A ft
    Seal Material _____

M   _Christy Box with locking well cap with lock_

Note: Depths measured from initial ground surface.

WELL NO.

Figure 1. Example of monitoring well detail.

sand pack, and borehole wall to flush and lift sediment into suspension for removal. At times a surge block may be used, followed by a bailer. If abundant silt or sand is present, a jet may be used to lift the material that surging and bailing cannot move and to aid in clearing the well screen. Jetting combines both pumping and air or water jetting. The high-speed air or water jet lifts the material out of the

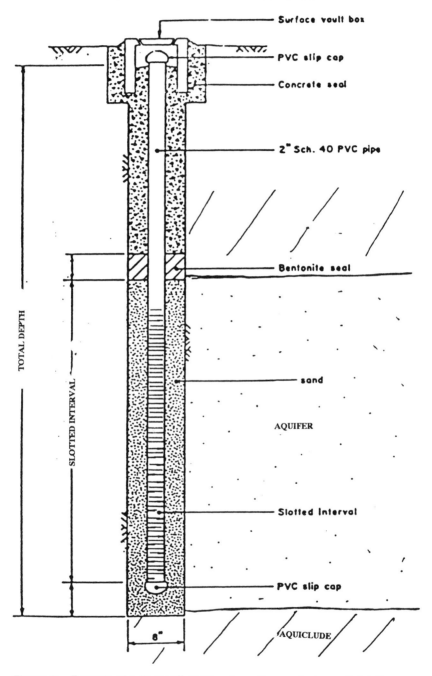

Figure 2.   Example of a "typical" shallow A aquifer monitoring well detail.

well, to be replaced by aquifer water. The well is jetted in discrete intervals. Jetting is effective for small-diameter wells.

The thoroughness of well development is critical, and the onsite geologist or hydrogeologist must carefully examine the discharged water to decide when development is sufficient (Peterson, 1990). Although water clarity guidelines are available, well development depends upon the well, the formation, and the consultant's field experience. The authors' experience has been that the degree of turbidity is dependent on the amount and completeness of development and formation texture. If development is complete, the drilling mud and slough should be removed and clear water is observed. All wells tend to yield some amount of sand and silt over time, even if the formation is relatively coarse-grained and the well design is well thought out. If the monitoring wells are in very low permeability formations, the development may finally yield water that is slightly turbid, which no degree of development can remove. Consequently, periodic well cleaning and redevelopment is needed to keep the well performing at optimum condition. If a well is sampled but not periodically cleaned, the well may silt in and inhibit aquifer water entry. It may even cut off water from the aquifer and become entirely plugged, rendering it useless for water sample collection.

WELL COMPLETIONS FOR DIFFERENT MONITORING SITUATIONS

The aforementioned example well installation is similar to that which may be used to monitor an underground tank or shallow groundwater in the uppermost shallow (or first-encountered A aquifer) water-bearing stratum. However, monitoring wells have to be placed in various situations to search for different potentiometric surfaces and stratified contaminant flow. Several different completion techniques can provide the needed coverage and include the B aquifer (or aquifer underlying A) completion, nested completions, and well clusters.

The B aquifer completion involves constructing a well through an upper contaminated aquifer (A) to the next underlying aquifer (B). Since the contamination must be kept in the A aquifer, a phased drilling approach must be used. First the upper and lower contacts of the A aquifer must be identified in a pilot boring to verify the depth to the aquitard underlying the A aquifer. The hydrogeologist then

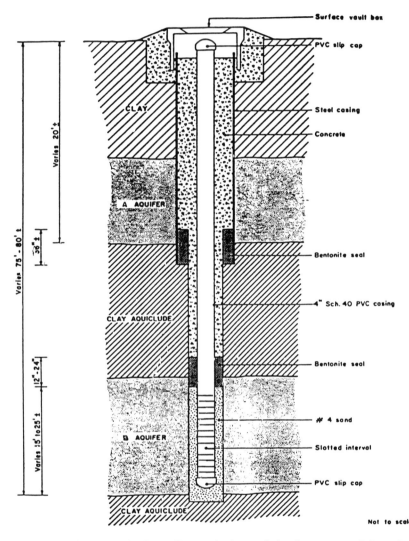

Figure 3.   Example of a B aquifer monitoring well detail constructed through a contaminated overlying aquifer.

enlarges the borehole to the aquitard and slightly into it. Extreme care must be used not to fully penetrate the aquitard since this provides natural containment for the upper contaminants. The larger diameter hole will allow a "telescoping" of the borehole with depth so that the upper aquifer can be cased off with a conductor casing. The conductor is lowered into the hole and pressed into the aquitard about a foot

Single Borehole Well Nest

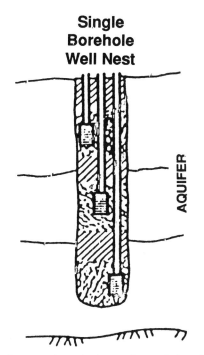

Figure 4. Example of a nested monitoring well completion. *Source:* EPA (1986).

to make good contact between conductor casing and aquitard clay. At this point the conductor may be cemented into place, and the drilling muck bailed from the hole. The hole is then flushed with clean water to remove contaminated mud and drill slough. Once completed, and after allowing the cement to set (typically 24 hr), drilling continues through the aquitard and into the second aquifer, where sampling and well completion occurs (see Figure 3).

A nested completion usually involves two or more monitoring wells constructed in the same borehole. Given time, money, and geologic constraints, this may be a faster way to complete wells and provide discrete monitoring from one borehole (see Figure 4). However, the complexities of construction are difficult, and again extreme care must be used to seal between screened intervals, especially if one aquifer is contaminated. Otherwise contaminated water may circulate to other wells, cross-contaminating aquifer zones. Another problem is borehole decay and erosion from utilizing one borehole for a long period of time. Boreholes will cave and erode (depending on the geology, fluid circulation, the duration the borehole is open, etc.),

Multiple Borehole Well Nest

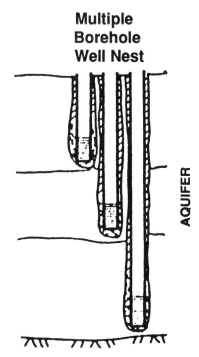

Figure 5.   Example of a cluster monitoring well completion. *Source*: EPA (1986).

and redrilling may be necessary. The amount of time spent in keeping the borehole open for construction may become prohibitively long in terms of budget and time. Although these completions can be useful for potentiometric data collection in one aquifer, monitoring well construction can be very difficult. Unless there is a compelling reason for use, the authors' opinion is that the nested completion should be abandoned in favor of using monitor well clusters.

Well clusters are individual wells constructed in close proximity to each other (usually within 5 to 15 ft). The wells are constructed as in the initial example, but each well screen penetrates a different depth. Thus, the wells monitor discrete depths within a single aquifer to observe contaminant stratification effects (see Figure 5). Although clusters are typically installed in groups of three, four, or five, an initial borehole or pilot hole may be drilled, and the stratigraphic and sampling information collected from that hole only. If the geology is determined to not vary significantly within the anticipated cluster area, then the wells are constructed in separate boreholes that are drilled to the design depth and constructed. The cost may be slightly more for drilling time, but the advantage in completion (and data) quality compensates for any cost differential. Additionally, the regulating agencies appear to be favoring the cluster installation for data and monitoring. When sites require extensive monitoring, this seems to be the most used approach and probably yields the best data by using discrete sampling points.

CHAPTER 4

Groundwater Monitoring Well Sampling

INTRODUCTION

Groundwater monitoring well sampling is one of the most important tasks performed during a contamination investigation. Together, the exploratory borehole log and the water quality information provide the two basic sources of subsurface data. Monitoring well sampling provides the groundwater geochemical and contaminant chemistry information for the problem under consideration. Many times, the investigator will be dealing with contaminants at the parts per million (ppm) or parts per billion (ppb) concentration level. Consequently, the possibilities of errors in data collection are enlarged, and extreme care and quality control must be used when obtaining samples.

The consultant is often asked to ascertain whether or not a contamination problem is present, and the chemical data will dictate the presence, nature, and type of problem. When the information is reviewed by the regulating agency, the presence of one type of contaminant, even at very low concentrations, may initiate a costly investigation and remediation. Hence, after properly constructing the monitoring well, the water sample collection procedures must be carefully thought out and executed to obtain reliable data. Finally, site specificity cannot be overemphasized. The aquifer chemistry may be very subtle or can change seasonally or throughout project life, and a large body of data will be required over time to observe the variations. The sampling procedures must be able to accurately define aquifer hydrogeochemistry.

SAMPLING PLANS AND PROTOCOL

Usually a sampling plan and procedures, commonly called the sampling protocol, are prepared prior to field investigation. A general sampling plan may take many forms and is used by consulting firms as their specific internal procedure document. Other sampling plans are written for specific projects or specific contaminant types. All plans are usually derived from the government guidance and amended by local agencies. At times the investigation and sampling plans may be reviewed by regulators, clients, and other involved parties, and it may take months, or even years, before a procedure consensus is reached. It is important to have sampling procedures that are similar to industry or regulated standards and provide the needed documentation of collection and handling from the field to the analytical laboratory. Regardless of project size or complexity, the level of care must be the same when collecting any subsurface data since it may be legally challenged at some future time.

Numerous agency-prepared groundwater sampling guidance documents are available for use in designing and writing sampling plans (for example, EPA, 1985a, 1986a, 1987b; State of California, 1986; Nielson and Johnson, 1990). The basic plan should provide for and collect all the related information and define procedures employed, with the data logged and managed through proper paperwork, including "chain-of-custody" and data collection forms. The information will ultimately create a project database to monitor changes in groundwater elevation, flow direction, and chemical evolution. This data would also be used to ascertain the contaminant extent and to design site remediation procedures.

SELECTION OF ANALYTICAL LABORATORY

The selection procedure of the chemical analytical laboratory should be as rigorous as the sampling plan preparation. The data collected from the site will be a function of the chemical analysis as well as sample collection, so the laboratory must be able to perform the required analyses according to approved analytical techniques. In some states, analytical laboratories are state-certified for types of analytical procedures, and local regulatory agencies may require that certification be a portion of the sampling plan. The use of EPA methods of chemical analysis are always required. Also a laboratory may have a specialty analyses set such as, for example, pesticides.

At times the laboratory nearest the site may not be able to perform the needed analysis, and the samples must be sent to another town or another state. Contingencies for rapid shipment and handling must be a part of laboratory service, if needed, since samples have a limited storage time prior to actual analysis. The laboratory must be able to document in-house quality control and assurance for sample handling and analysis and to provide the record-keeping and raw data files to check the veracity of samples when data reliability is questioned. Finally, the laboratory chosen should be used for the duration of the project, so differences in quality control resulting from changing laboratories do not become a problem. At times, different laboratories must be used, and since no two labs are alike, data questions may arise due to slightly different procedures, instrumentation, and information recording (data expression).

QUALITY ASSURANCE AND QUALITY CONTROL OBJECTIVES

Quality assurance and quality control (QA/QC) procedures are established to obtain field data for water quality in an accurate, precise, and complete manner so that information is accurate and representative of actual field conditions. This involves both field and laboratory procedures for internal and external checks on data collection. The data collected are usually checked for accuracy, precision, completeness, comparability, and representativeness. Definitions for these terms have been set forth in guidance documents (for example, EPA, 1986a), and as applied to groundwater quality sampling are as follows:

Accuracy: the degree of agreement of a measurement with an accepted, referenced, or true value

Precision: a measure of agreement among individual measurements under similar conditions, usually expressed in terms of standard deviation

Completeness: the amount of valid data obtained from a measurement system compared to the amount that was expected to meet the project data goals

Comparability: the confidence with which one data set can be compared to another

Representativeness: a sample or group of samples that reflects the characteristics of the media or water quality at the sampling point; also how well the sampling point represents the actual parameter variations under consideration

These QA/QC objectives, as well as the following suggested sampling procedures, have been extensively documented and presented in government publications and guidance books and in some codes and regulations of various federal, state, and local agencies.

PREFIELD PREPARATION

The sampler must prepare for a sampling session by decontaminating and calibrating the equipment. Today, much sampling equipment is built of inert materials, such as stainless steel and Teflon, to minimize potential contaminants from equipment. Equipment cleaning is accomplished by washing the bailers, pumps, hoses, lines, and related gear with trisodium phosphate or alconox-type soaps, followed by a distilled or deionized water rinse, and occasionally a solvent rinse, as the protocol directs. Steam cleaning of equipment may also be used as required. If cleaning and decontaminating cannot satisfactorily clean the equipment, then it should not be used and should be replaced. Equipment used to measure pH, depth to water, conductivity, dissolved oxygen, and so on should be checked daily for calibration accuracy according to the individual manufacturer specifications. Since this equipment yields basic general information on ambient aquifer conditions for each sampling event, proper calibration is vital.

SAMPLE BIAS

Sample bias may be difficult to determine, given the large potential for something to go wrong. Sampling technicians' training and techniques may vary widely between individuals or companies. Different sampling methods have their own advantages and disadvantages. The method should be appropriate for contaminant type (for example, volatile organic chemical, trace elements, pesticides). The site-specific sampling method and procedure protocol must be carefully followed so that sampling bias will not affect the reproducibility of the results.

In other words, one does not want field sampling error to affect laboratory results.

Sampling bias could result from a number of possible sources, including improper purging of the well, sampling wells in low-permeability aquifers, improper cleaning of equipment, and sampling from first contaminated wells and then clean wells. The sampling protocol should try to take these potential problems into account so that they can be avoided. For example, if oily contaminants are present, steel sampling gear may be an option since plastics and Teflon may be difficult to clean. If sampling equipment cannot be cleaned, then it should not be used and should be replaced with acceptably decontaminated equipment.

Collecting and using sample blanks serves as a check on possible laboratory error or bias. The blank checks on both the laboratory and sampler for trace amounts of contaminant and possible lab error. Thus, the blanks are very useful for checking on overall data accuracy. Usually, a certain percentage of blanks are factored into the analytical program, determined on a rate-specific basis or generally ranging from 5 to 10% of the total number of samples. Additional random blanks may be run from time to time on the sampler for equipment cleanliness, and spiked blanks (blanks containing a known quantity of contaminant) can be sent to the laboratory to check on the accuracy of the actual analysis. The most common types of blanks used are the following:

Trip blanks are used for purgeable organic compounds only. They are sent to the project site and travel with the collected samples. Trip blanks are not opened and are returned and analyzed with the project samples.

Field blanks are prepared in the field with organic-free water. These samples accompany the project samples to the laboratory and are analyzed for specific chemical parameters unique to the site at which they were prepared.

Duplicates are collected as "second samples" from a selected well and/or project site. They are collected as either split samples (collected from the same bailer volume or pumping discharge) or as second-run samples (separate bailer volumes or different pumping discharges) from the same well.

Equipment blanks are collected from the field equipment rinsate as a check for decontamination thoroughness.

EXAMPLE OF FIELD PROTOCOL FOR SAMPLING A MONITORING WELL

Initial Steps

The sampling data forms are prepared; the well sampling plan usually samples from the least to most contaminated well. Again, this is done to minimize the potential of cross-contaminating wells. If the site is to have very long-term sampling for required monitoring, the most prudent course is to install dedicated sampling devices in each well so that the potential for cross-contaminating wells and equipment contamination is further minimized. Whether the system is composed of bladder pumps, submersible pumps, or some other system, the cost of installation and maintenance is more than offset by reducing the possibility of collecting suspect data, which may initiate redundant sampling or additional remedial work (Table 1).

Upon arriving at the monitoring well, the sampler should have clean equipment and properly prepared and preserved sample containers. The sampler will begin data collection by recording the project number, date, time, and site conditions, such as weather, well security, signs of vandalism, and any other required information. Once the well is uncapped, the depth to water is measured to the nearest 0.01 ft and logged for the water elevation calculation. This measurement is taken at the project datum survey tick mark. At this time the well may also be checked for the presence of a separate contaminant phase floating on the water. The apparent thickness of product may be checked with a bailer or water-finding pastes. Optical electronic probes may be used to simultaneously measure depth to "product" and depth to water. Where there is floating product, the apparent product thickness may be measured with bailers.

Monitoring Well Purging

Following these initial steps, the well is then purged prior to collecting the actual sample. This is very important since the geochemistry of the water in the well casing may change due to stagnant conditions and exposure to the atmosphere. The sampler should calculate the required volume ($v = \pi r^2 h$) that must be removed to draw aquifer water into the well. Hence, the sampler must know the depth and diameter of the well casing and borehole (to take into account the sand pack). Most regulating agencies recognize three to four borehole

Table 1.   Generalized Groundwater Sampling Protocol

Step	Goal	Recommendations
Hydrologic measurements	Establish nonpumping water level.	Measure the water level to ± 0.3 cm (± 0.01 ft).
Well purging	Remove or isolate stagnant H_2O, which would otherwise bias representative sample.	Pump water until well purging parameters (e.g., pH, T, Ω^{-1}, Eh) stabilize to ± 10% over at least two successive well volumes pumped.
Sample collection	Collect samples at land surface or in well bore with minimal disturbance of sample chemistry.	Pumping rates should be limited to ~ 100 mL/min for volatile organics and gas-sensitive parameters.
Filtration/preservation	Filtration permits determination of soluble constituents and is a form of preservation. It should be done in the field as soon as possible after collection.	Filter trace metals, inorganic anions/cations, alkalinity. Do not filter TOC, TOX, volatile organic compound samples; other organic compound samples only when required.
Field determinations	Field analyses of samples will effectively avoid bias in determining parameters/constituents that do not store well (e.g., gases, alkalinity, pH).	Samples for determining gases, alkalinity, and pH should be analyzed in the field if at all possible.
Field blanks/standards	These blanks and standards will permit the correction of analytical results for changes that may occur after sample collection: preservation, storage, and transport.	At least one blank and one standard for each sensitive parameter should be made up in the field on each day of sampling. Spiked samples are also recommended for good QA/QC.
Sample storage transport	Refrigerate and protect samples to minimize their chemical alteration prior to analysis.	Observe maximum sample holding or storage periods recommended by the EPA. Documentation of actual holding periods should be carefully performed.

Source: EPA (1987).

volumes as a sufficient quantity for removal, although additional volumes may be required by local agencies or project-specific needs. However, there is no set criteria for using a set number of purge volumes, and specified purge volume numbers are misleading and could yield questionable data (EPA, 1985a).

Schmidt (1982) reports that, based on experience with monitoring wells in the southwestern United States, 30 to 60 min of pumping at rates of 20 to 50 gpm are needed before chemical parameters stabilize in highly transmissive alluvial aquifers. This may be an extreme case for purging. Behnke (1990) states that high-rate purging may squeeze clays, which may cause changes in inorganic water geochemistry. Hence, low purge rates are recommended to maintain stable pH and electric conductivities. Wells should be purged according to the site-specific hydraulic well performance for "representative" sample collection (Gibs and Imbrigiotta, 1990). Unless minimal purge volumes are removed, only casing storage or "stagnant" water will be sampled, which is not representative of the aquifer (Driscoll, 1986).

While the purge volumes are being removed, the basic physical parameters of pH, conductivity, temperature, and dissolved oxygen are monitored to help judge when the aquifer water is entering the well (EPA, 1985a). Usually these parameters fluctuate during purging, finally settling to a steady value or range of values. For example, typical parameter stabilization ranges may be tenths of degrees Fahrenheit, tenths of pH units, and tens or hundreds of conductivity units. Once relatively constant values are observed, the sampler may sample the aquifer water entering the well. If the well recharges quickly, then the sample collection may commence. If the well has dewatered somewhat, then the well should be allowed to recharge, and commonly the sample is collected upon 80% recovery to the originally measured static water level. This information should also be used to build a database of well parameters for comparison with future sampling rounds, providing information on long-term fluctuations in aquifer general water quality and well performance.

Groundwater Sample Collection

The groundwater sample may now be collected. Depending on the sampling technique, the sample is either pumped or bailed to the surface. Care must be taken so as not to agitate the sample to cause volatilization of low vapor pressure contaminants (thus, sample cavitation must be minimized while retrieving the sample) (see Figure 1).

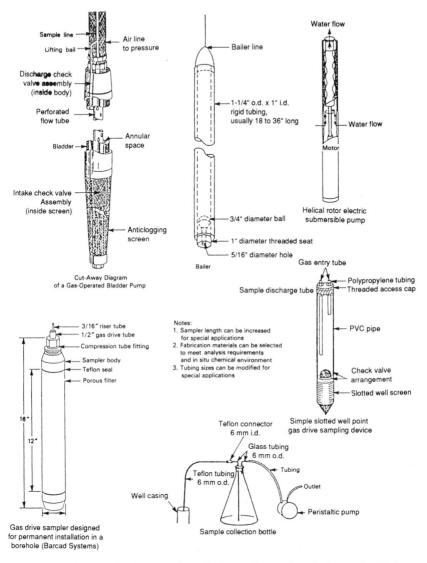

Figure 1. Schematic diagram of monitoring well sampling devices, after Neilson and Yeates (1985) in EPA (1987c).

The sample is carefully poured with minimum headspace into the appropriate sample bottle and immediately capped. The sample bottle is labeled with the date, well number, sampler name, and other information if needed, and placed in a refrigerated container. The sample may be turbid even after the purge (most wells will pump a little sand and silt almost all the time). A filtering step may be desired when analyzing for metals, or if the turbidity has been known to interfere with analysis. The sampler should review the protocol and data with the hydrogeologist or geologist to determine when filtering is needed as well as to adapt the sample protocol to observed site conditions.

All samples are logged onto chain-of-custody forms so that a complete sample-handling record is kept to ensure proper transport and care. Each person responsible for the samples must sign, with date and time, for sample custody. Since proper handling and procedures are vitally important, this becomes a legal document so that all samples can be tracked through the project life for regulatory needs and any legal challenges to sample data. Once complete, the sampler decontaminates the equipment, recaps and locks the wellhead, and moves to the next well.

SAMPLING LOW-PERMEABILITY OR SLOWLY RECHARGING AQUIFERS

Sampling wells in low-permeability environments may cause peculiar problems. Typically, these aquifers yield low quantities of water, and the fine-grained material (clay) causes the well water to be turbid. When the sampler performs the well purge, the well may dewater and not recharge for minutes, hours, or even days. Similar problems may arise when sampling from aquifers during periods of drought. Sample protocol must be adapted for the low recovery of the monitoring wells and samples collected as the hydrogeology dictates, perhaps requiring departure from generally accepted sampling procedures. It may even call into question the definition of "monitoribility" and require interpretation of long-duration flow situations, given long sample retrieval intervals under regulated sites (Marbury and Brazie, 1988). Hence, the well performance and geology create problems for traditional concepts of monitoring set forth by regulation. Needless to say, negotiation with interested parties will be required for a consensus on sampling methodology and schedule.

The sampler may retain the purge water if the well yield is extremely low, and under special circumstances may use this for the sample analysis. If the well recharges over a 24-hr time period, then the sampler may return and collect the sample. Several samples may need to be collected, and a range of chemical quality used for the site background or to mark contaminant presence. It is probable that the sample will be turbid given formation conditions, and sample filtering may be necessary. Finally, if the well does not recharge, sampling is canceled for that time interval, and the sampler must wait for the aquifer recharge and the next scheduled sample collection date.

The slowly recharging aquifer can present problems of accuracy of aquifer chemical conditions. Volatilization can occur from water trickling through the sand pack and casing. McAlony and Barker (1987) report that volatile compound losses of 10% may result in 5 min due to recharge water trickling through dewatered sand pack. It may not be possible to collect "representative" samples in the currently understood sense, and the aquifer chemistry may not be comprehended even after numerous sampling rounds.

These problems have to be mediated between the client, consultant, and regulator to agree upon the sampling protocol and criteria for a specific site. Since the site hydrogeology ultimately dictates the nature of the groundwater occurrence and chemistry, flexible protocols may be needed. In this way, the maximum amount of data can be collected to study a site where groundwater information is difficult to acquire. Interpretation of chemical analyses from samples collected under these conditions should be carefully reviewed. The data accuracy confidence may be low, especially when ascertaining the presence or absence of contaminants that were thought to occur at very low concentrations.

LIQUID SAMPLING IN VADOSE (NONSATURATED) ENVIRONMENTS

At times, collection of liquid samples from nonsaturated environments is needed in landfill or land treatment sites to track possible contaminant movement through the vadose zone toward the aquifer (EPA, 1985a). The technology for this type of sample collection has been available for years and was developed by agricultural science to study water movement in crop root zones. The sampler is the pressure vacuum lysimeter which operates on the principle of using soil suction to cause movement of liquid adhered to soil particles into the sampler. A complete review of vadose sampling and soil moisture detection has been complied by Wilson (1981).

The lysimeter is a capped tube with a permeable ceramic cup at the bottom, through which the sample is collected. The ceramic cup is

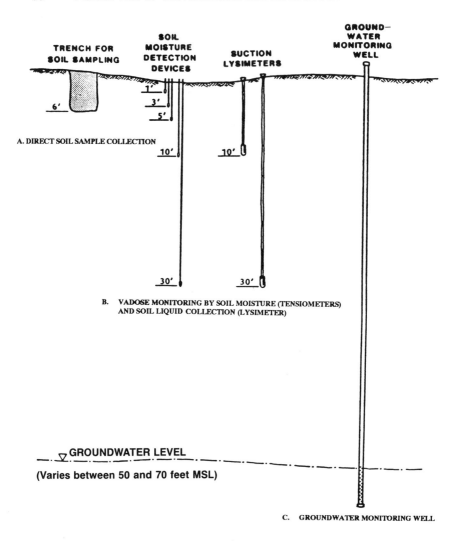

Figure 2. A combined monitoring system.

manufactured to a certain pore opening size to allow liquid entry. These samplers are inexpensive and versatile, and can be used up to several hundred feet deep. Oily or chemical contaminants tend to clog or alter the permeability of the ceramic, and may render it useless for sample collection. Another problem is that ceramic material may leach heavy metals from ceramic bulb or Teflon cups and bias the sample. Lysimeters are relatively cheap, and when carefully installed they can provide chemical data from soil liquid that would otherwise be unavailable in the unsaturated zone.

Lysimeter installation usually involves collection of soil moisture data so that an estimate of available liquid can be made. Soil samples may also be collected to ascertain indigenous chemistry or contaminant presence and soil moisture, which help to calibrate conditions at the start of the monitoring program. In some cases, tensiometers or other moisture-measuring devices are used to measure increased moisture, a signal that soil liquid sampling should start. The lysimeter unit is installed in a borehole with the ceramic tip encased in a fine-grained porous material (such as very fine sand-blasting glass beads), which forms a hydraulic envelope about the cup to protect it from clogging. Once installed, pressure is applied to the unit, which induces a suction at the ceramic creating a pressure gradient to overcome moisture surface tension and cause moisture movement of the water films on soil particles. After a nominal time period (several to 24 hr), the pressure is released, and the sample that has accumulated in the tube is pumped to the surface and collected from the discharge line. Lysimeters can be used in low-moisture soils and can have long and useful service lives when properly installed and sampled (Merry and Palmer, 1985). A conceptual example of combined monitoring using soil sampling, lysimeters, and monitoring wells is shown in Figure 2.

CHAPTER 5

Regulatory Frameworks

INTRODUCTION

The field of groundwater quality and protection has become highly regulated over the past 20 years as environmental awareness grew in the scientific and general community. Hydrogeologists dealing with contaminant groundwater studies now should have a working knowledge of the law and regulations pertaining to groundwater, potable water quality, and aquifer protection. Related areas involve liability and ethics issues for hydrogeologists and other consultants. These and other issues are problems peripheral to the actual science of hydrogeology, but consultants should be aware of the possible effects on their investigations and field studies.

Obviously, the legal issues are book-length subjects in themselves. Legal and regulatory issues are covered and updated by books and seminars on a regular basis. The intent of this chapter is to give an overview of some applicable laws that hydrogeologists may encounter and how they must function within the legal and regulatory boundaries. This chapter is meant to introduce readers to the regulatory world in the most general sense, assuming that they will have to address regulatory issues and be knowledgeable of related legal issues during their investigations. Readers are referred to the U.S. Code of Federal Regulations, and applicable state and local regulations for specifics; they must deal with these regulations on a case-by-case basis.

Existing laws have done much to order and direct the environmental goals, but understanding them can be a bewildering task. The federal and state codes are the primary sources of the laws and regulations governing contamination. However, an excellent reference is Elliott (1987–1991), which presents a matrix and explanation of legislation by state. Currently, several states (California, Florida, Illinois, New Jersey, Ohio, Pennsylvania, and Texas) have been compiled by

Elliott and his associates, and the following discussion borrows heavily from his publication.

OVERVIEW OF FEDERAL LAW

Numerous laws and regulations have been passed in the 1970s and 1980s that deal with environmental issues and groundwater. Many federal laws have been passed (including the creation of the Environmental Protection Agency) that now govern or guide groundwater contamination investigations:

National Environmental Policy Act of 1970

Federal Water Pollution Control Act of 1972

Toxic Substances Control Act of 1976

Resource Conservation and Recovery Act (RCRA) of 1976

Clean Water Act of 1977

Surface Mining Control and Reclamation Act of 1977

Safe Drinking Water Act of 1979

Comprehensive Environmental Response, Compensation and Liability Act (CERCLA) of 1980

Hazardous and Solid Waste Amendments (HSWA) of 1984

Superfund Amendments and Reauthorization Act (SARA) of 1986

These laws established environmental protection and cleanup, as well as the definition of hazardous waste, hazardous waste transport, handling, cleanup, and disposal. Laws were also added to address different aspects of environmental problems. Interestingly, there is no federal groundwater protection statute (Patrick, Ford, and Quarles, 1987). Although federal laws do provide for groundwater protection, they tend to focus on a narrow range of polluting activities (Patrick, Ford, and Quarles, 1987). The Safe Drinking Water Act of 1974 provided EPA authority to promulgate primary and secondary drinking water standards to public water supply systems. EPA's Office of Groundwater Protection released guidelines for groundwater classification in 1986. That system consists of three general groundwater classes: I—special groundwater, II—groundwater currently and potentially a source of drinking water, and III—groundwater not a source of drinking water. This system is based upon drinking water as the highest beneficial use of the resource.

A complete legal review is beyond the scope of this book; readers are referred to the various applicable codes of regulations. However, to introduce the reader to the legal world, a highly generalized and brief review of some related federal and state regulations is presented since these are concerned with groundwater issues. It will also illustrate how complex the legal frameworks can become.

Resource Conservation and Recovery Act and Comprehensive Environmental Response, Compensation and Liability Act

The Resource Conservation and Recovery Act (RCRA) law intent, as summarized by Elliott (1987–1991), is to provide the "cradle to grave" regulation of hazardous wastes, and is actually constructed upon the Solid Waste Disposal Act of 1965, RCRA, the Hazardous and Solid Waste Amendments (HSWA) of 1984, and part of the Superfund Amendments and Reauthorization Act (SARA) of 1986. The portions most often encountered by hydrogeologists are contained in the Code of Federal Regulations (CFR) 40, Part 264, Subpart F, which provides the geologic and hydrogeologic investigation approaches, siting, and groundwater monitoring of hazardous waste disposal, storage, or treatment sites. These regulations outline the geologic information that federal regulators require for protection of aquifers underlying and near these sites, including seismic safety, engineering, geologic, and geotechnical engineering requirements, in addition to the groundwater aspects of facility siting.

Basically, the groundwater requirements should provide a knowledge of the subsurface, installation of properly designed and constructed monitoring wells, and understanding of site-specific background groundwater geochemistry and contaminant data. In this way, the facility geology and its relation to nearby drinking water sources can be utilized to protect those resources. A statistical analysis of the monitoring information and indigenous baseline water quality is compiled and used to ascertain whether significant changes have occurred in water quality at detection monitoring points. If so, then monitoring at a quality standard compliance point is established and maintained while the site is cleaned up. Once cleanup is finished or facility use ends, then a postclosure monitoring is performed for a considerable duration (usually 30 years) to ensure that the problem has been properly solved. Needless to say, these are long, complicated, and highly costly investigations, requiring many staff members and much effort on the part of the consultant and client, and usually negotiation with the regulators. Although federal law is phasing out landfills and disposal to land, these types of investigations will continue for some time to come.

A related law is the Comprehensive Environmental Response, Compensation and Liability Act (CERCLA) of 1980, or "Superfund," which was created to identify sites contaminated by release of hazardous materials and finance the remediation by "responsible parties" or from federal or state cleanup funds. CERCLA was amended by SARA, which strengthened Superfund. Although EPA implements this program, some elements may be controlled by the states, which often become the lead agency for a particular cleanup.

Basically, sites are initially identified to EPA by owners and regulatory agencies, including those already identified under RCRA. Next, the Superfund sets priorities for cleanup under the National Contingency Plan using a rating system. The third step is to identify the owners and clean up the sites. The agencies seek to identify the "potentially responsible parties" (PRP) who can be required to pay for the remedial investigation/feasibility investigation to determine the extent of contamination and provide the bench test and design of the treatment and cleanup, and the cleanup itself, directly or through reimbursement of federal expenditures. If no PRP is identified, then the Superfund finances these activities.

EPA and State Underground Storage Tank Programs

RCRA and CERCLA are used for industrial, military, and civilian sites to clean up the contamination problems. However, another scale of contamination problems exist that are now coming to the forefront—leaking underground storage tanks, which occur everywhere in the United States, most prominently at the corner service station. Large contamination problems can arise from leaking tanks, whether the problem arises from large tank farms or from one leak releasing a large quantity of contaminant, especially if classified hazardous. Consequently, many states have or will be implementing underground storage tank (UST) programs to deal with these problems, which can be from numerous diffused sources or a large point source. Some states, notably California, Florida, and New York, instituted UST programs in 1983-1984; most states are now responding to federal standards developed by EPA under HSWA requirements.

The UST approach differs somewhat from the RCRA approach—

the scale is usually smaller in investigation effort and sometimes financial costs; however, the level of care must be the same for both types of studies. If not, the smaller problem can easily grow to huge proportions, together with legal and liability problems rivaling RCRA-CERCLA.

EPA UST Program

The EPA UST program (40 CFR, Parts 280 and 281) was passed to provide for regulation of underground tanks nationwide. This program will work with the state and local agencies to detect leaks, prevent spills, monitor tanks, and clean up subsurface leakage. In addition, the tanks must be properly closed, and a financial responsibility section is included to address damage and cost of cleanup. While many tanks apparently fall under this program, many others do not — farm and residential tanks smaller than 1100 gal storing for noncommercial use, onsite heating oil tanks, tanks on or above the floor of underground areas, septic tanks for storm and wastewater, flow-through process tanks, tanks of 110 gal or less, and emergency spill and overflow tanks.

Other sections of the regulations discuss overflow and corrosion prevention and detection, soil and groundwater monitoring, secondary contamination of tanks, interstitial monitoring for chemical storage tanks, and notification of agencies. There is guidance for monitoring well placement, allowing some flexibility for well placement and design, especially for immiscible (floater) contaminants. As these regulations are implemented and evolve, changes and amendments will occur as the program goes into effect.

Selected State and Local Regulations

Several other states have passed UST laws and regulations in the past several years, including very extensive programs in New Jersey and Florida. California passed a comprehensive tank law in 1983, and several counties had passed underground tank ordinances prior to the statewide law. In all cases the intent in all states is similar — to provide guidance on basic requirements for monitoring, construction of monitoring wells, soil and groundwater sampling, and periodic monitoring of the subsurface tanks and piping installation.

Example of Development of Local Regulations

Discovery of leaks from subsurface solvent and fuel storage tanks in the Silicon Valley, located in Santa Clara County at the southern end of the San Francisco Bay, created concern about hazardous materials. The Hazardous Materials Storage Ordinance (HMSO) was developed in Santa Clara County in 1982–1983 by a task force including participants from local government. The Santa Clara Fire Chiefs' Association sponsored the effort since the firemen needed to know the location of aboveground hazardous material storage. Other groups joined the effort since they were concerned about leaks of material from underground storage tanks. In 1983, Santa Clara County and its fifteen constituent cities began to implement HMSO. As of 1989, there were 100 local UST programs in California, with 57 run mostly by county health agencies and 43 run in cities mostly through fire departments (Elliott, 1987–1991).

The Santa Clara Valley Water District, a local water district, implemented underground tank groundwater monitoring guidance in 1983 in response to the numerous tank leaks that had occurred from both fuel and chemical storage. This ordinance provided for installing monitoring wells, instituting periodic monitoring for both water and soil vapor, and analyzing soil samples from the vicinity of the tank or tank complex. Soil and groundwater sampling protocols were established. Permits were required so that monitoring well locations were public knowledge, as well as to review the basic construction specifications required by the ordinance. This documented the local site geology and provided a preliminary estimate of soil and water quality and leakage extent.

Thus, layers of regulations may exist at one site — city or county, water district, state, and possibly federal, depending on the case. Hydrogeologists must therefore review the regulations and agencies with which they will be involved. This has a direct bearing on the completeness of the work required to address the needed guidance and reporting format. If the work is planned solely to address the specific regulations, then the work may be considered incomplete. If monitoring points are not sampled according to an accepted protocol, collected data could be deemed invalid. It is always the responsibility of the supervising technical professional to be conversant with the laws and regulations pertaining to that location and site.

AGENCY INVOLVEMENT AND NEGOTIATION WITH THE CONSULTANT

The regulating agencies may not always have the "answer" for the consulting hydrogeologist if the law is unspecific. For example, many groundwater regulations exist, but soil contaminants in the vadose zone, above groundwater occurrence, are not regulated as specifically. Other contaminants may be involved for which regulated standards do not exist or toxicological data are incomplete. Therefore, although these contaminants may need to be removed, cleanup guidelines may utilize concentrations of contaminants, or even other regulations, not originally conceived for these particular cleanup purposes. At times, soil samples may reveal contamination, but the meaning of the chemical concentrations will be unclear. In California, guidance may include transport regulations, which determine if the constituent is present at levels deemed to be hazardous, or ascertain the implied threat to groundwater, using EPA guides for discharging wastes to landfills in concentrations 100 times greater than the water quality goal.

Agencies always set the ultimate sampling frequency, types of constituents for which analyses will be done, and the cleanup standard (how clean the site needs to be). For example, a site may be required to be cleaned to 1 ppb of chemical X and monitored for some duration on a quarterly basis. These would be site-specific standards. Since approaches to regulation application are evolving and the toxicological knowledge base is growing, including contaminant transport, final disposition cleanup goals will change. Negotiation of the site problems may involve policy of investigation approaches and changes in how chemical analyses are performed. These are difficult problems, and no easy solutions are available. However, the difficulty may be eased if the consultant has a working knowledge of the applicable regulations, has conducted a complete and thorough investigation, and has experience negotiating with clients and agencies.

Since the consultant is dealing with laws and supervising agencies, the investigation paper flow must be sent to these people. The consulting hydrogeologist is acting on the client's behalf in presenting site information and plan development, either on paper or orally, to the regulating agencies. The consultant will have to send written plans, rationales, sampling protocol, periodic monitoring reports, and other documents for agency review to support the consultant's case and interpretation of the collected data. Also, the consultant will need to meet with the case officer assigned to the site. This can involve negotiation of investigation approaches, data validity, monitoring system development, and ultimate site cleanup plans. The subsequent remediation plan negotiation is predictably difficult since the extent of cleanup will dictate how much money will be required. Obviously, the "how clean is clean?" debate can become vigorous. Since the agency will decide how far the cleanup will proceed and set the chemical concentration cleanup levels, the consultant's input mainly involves the site hydrogeochemistry. The burden of proof will rest with the client and consultant so that the regulator understands the situation well enough to establish the cleanup goal that is realistic for the site. These are difficult problems to solve, and all parties desire the same end — environmental compliance and protection.

PRELIMINARY SITE ASSESSMENTS—AN EXAMPLE OF NEGOTIATION WITH REGULATORS

A growing field of environmental work is preliminary site assessments, commonly referred to as PSAs, or sometimes called "environmental due diligence." These types of investigations are done to ascertain whether a site has an environmental problem (in this case, subsurface contamination). Since property owners, banks, or other involved parties may become liable for environmental compliance or cleanup, these studies are increasingly involving the site history and contaminant potential. The need for future environmental compliance may or may not be required by a regulating agency regardless of property ownership. In order to avoid potential liability, clients will need to prove that they did not cause the problem or that the problem came from an offsite source.

Palmer and Elliott (1988) have suggested a general approach for this type of problem. When working with the agencies, the consulting hydrogeologist should supply the needed information to the regulators so that they can evaluate site conditions. If some information is not forwarded to them, or incomplete work is performed, the position you wish to present may not be acceptable. All records, field data, and chemical data should be completely documented, and copies filed with the agencies so that a paper trail exists in their files. Unreasonable demands and confrontational meetings tend only to polarize individuals and hinder progress for both sides. These projects will have substantial costs in time and money, regardless of size. The best approach is to have the material in hand, together with historical files to document and support your position. Good lines of communication are needed with regulators that accurately state the site conditions and needs in the context of applicable regulations. Negotiation should be firm but amicable. Two case histories are given from Palmer and Elliott (1988) that illustrate this approach.

Case 1

A convenience store possesses onsite underground gasoline storage tanks which are monitored by UST leak detectors and one monitoring well. A sudden appearance of gasoline in the single monitoring well indicates product is moving under the site. The tank leak detectors show no leak. The owner elects to install two more monitoring wells on his upgradient property line and discovers gasoline migrating onsite from an upgradient source. He monitors this quarterly. Six months later, state and local agencies inform the owner and several other nearby subsurface tank owners that they may be named as responsible parties for a large gasoline plume in the vicinity.

Our store owner sent the monitoring well installation reports and quarterly monitoring to the state, so they were in state files months before the state started looking for responsible parties. This evidence of two upgradient wells, and quarterly monitoring of all three wells, shows he is not responsible. The state does not name our owner as a responsible party, but does request that he share monitoring information.

By installing two wells and keeping abreast of the problem, the store owner saved himself from being entangled in the responsible party battle, and cooperated with the agencies. At a cost of about $12,000 dollars over one and one-half years, he saved himself from being ordered to do a $50,000 investigation over a three-month period — and the future grief from being involved in a cleanup problem that he did not create.

Case 2

A development company (buyer) wants to buy a portion of a valuable property. Although the site does not appear to have had hazardous materials on it, the buyer elects to perform an environmental reconnaissance investigation by installing monitoring wells and collecting soil and groundwater samples for chemical analysis. Samples from the wells reveal slight contamination beneath the site by industrial solvents in concentrations that exceed state action levels. Monitoring wells across the property line, installed by another consultant for the owner (seller), confirm the contamination, but it turns out that the owner does not have a history of use of these chemicals. The state agency would want "proof" that the contamination was not caused by the current owner since the state surmised a history of solvent use on or near the subject property. Otherwise the state may want the current or future owners to become involved in a future investigation and cleanup.

Since two consultants are involved, one for the buyer and one for the seller, additional investigation is done to verify that contaminants exist, confirm which way water is moving, and localize history of spills in the region. The additional surficial soil sampling shows that contaminants are not present and so could not have migrated through the vadose zone to groundwater. This eliminates an onsite source. Several large industrial solvent spills had occurred upgradient and were moving in the direction of the property. The limited data from monitoring wells and other publicly available reports on those large spills indicate that the plumes may have moved to the site, or the plume edge was in the vicinity of the property.

Here the state does not render a "final" option as to responsibility, but can require future work. But the proof gathered and shared by the two consultants, collected from numerous sources, shows that (1) the site soil was not the source of the contaminants, (2) large upgradient solvent plumes had occurred and were moving toward the subject property, and (3) the initial presence of solvents and sources were traced, if not to a point of origin, at least to several possible sources, which did not involve the current owner or prospective buyer. The approach has been to collect the data and construct an argument that takes into account the types of information that the state would request to evaluate contaminant sources as if an onsite investigation were required. The work performed should meet the requirements of the state if a future challenge is made. Since the sources of contamination are traceable to some offsite point, the likelihood of being named in a cleanup seems remote, but not certain. Both owner and potential buyer now must resolve potential risks and fine points of the sale given current hydrogeochemical knowledge.

SUMMARY

The consultant must have a good working knowledge of federal, state, and local regulations regarding the site under consideration. At times a layering of regulation may exist, presenting the need to interact with several agencies. The burden of proof will usually be on the clients (and their consultants) to show the extent of contamination and negotiate remediation. Typically, there will negotiation with agency people, and complete and up-to-date information is needed for agency review. Negotiation should be firm and amicable for the work necessary to get site-specific answers. The consultant should obtain information that anticipates regulator questions and provides answers within the context of applicable regulations.

CHAPTER 6

Introduction to General Groundwater Geochemistry

INTRODUCTION

This chapter considers geochemistry and laboratory analysis as related to applied contaminant hydrogeology. A vast body of literature exists on groundwater geochemistry, a thorough review of which is beyond the scope of this chapter. Any site has an indigenous natural geochemistry or chemical fingerprint which has to be quantified at the beginning of the study. Also, investigators will have to evaluate the anthropogenic contaminant geochemistry for which they are searching. We will discuss two aspects of groundwater quality: first, general geochemical parameters that have regulated criteria for drinking water and for the protection of drinking water aquifers; and second, laboratory selection and analyses.

Naturally occurring groundwater has a geochemical variability caused by natural processes, including the flow of the groundwater, the formations through which flow occurs, chemical changes resulting from annual flow fluctuations recharge sources, and mixing with other groundwater sources having differing chemistry. Elements enter or leave the system, or form compounds within it (see Figure 1) (Toth, 1984). Groundwater quality depends on the substances dissolved in the water and the chemical behavior of the compounds in the water imparted by the indigenous geology through which it flows. Consequently, groundwater quality is changeable as water moves through the aquifer. Changes may also be imparted, albeit very slowly, from leakage through aquitards or aquicludes (see Figure 1) (Toth, 1984).

It is very important to quantify the geochemistry of the groundwater at the site under consideration so that the baseline quality is established at the beginning of the study. This includes both the soil and groundwater since recharge through the soil or sediment cover will generally chemically impact upon the water in the aquifer. As

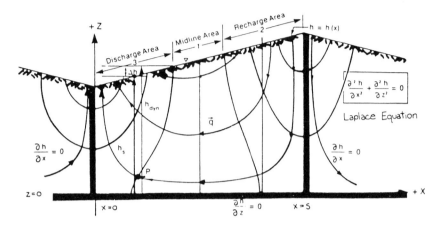

Ground Water Chemistry and
Hydraulic Regimes of Flow Systems

Recharge areas

Conditions: rain water; low TDS; high CO_2; low T; gradp → $\vec{q}$; cross formational $\vec{q}$

Processes: dissolution, hydration, oxidation; attack by acids; base exchange

Consequences: Dominant species — Ca, Mg, HCO_3, CO_3, SO_4
diverse rock types → diverse constituents
rapid increase in TDS

Midline areas

Conditions: source water moderately changed; p ≈ hydrostatic; $\dot{\rho}_k$ ≈ 0; T ≈ const; low free CO_2; little cross formational $\vec{q}$

Processes: dissolution, precipitation, sulfate reduction, base exchange

Consequences: Dominant species — Na, Ca, Mg, HCO_3, SO_4, Cl; gradual increase in TDS

Discharge areas

Conditions: highly mineralized source water gradp and gradT opposite to $\vec{q}$; cross formational flow; mixing with descending freshwater.

Processes: precipitation; reduction; membrane filtration

Consequences: high TDS possibly decreasing upward; dominant species — SO_4, Cl, Na

General Changes in Direction of Flow

TDS : Increase

$\dfrac{SO_4}{Cl}$: Decrease (SO_4 reduction; higher solubility of Cl)

$\dfrac{SO_4}{HCO_3}$: Increase (depletion of CO_2)

$\dfrac{Ca}{Na}$: Decrease (no Ca added due to depletion of CO_2; exchange of Ca for Na)

$\dfrac{Ca}{Mg}$: Decrease (no Ca added; Mg SO_4 more soluble than Ca SO_4)

(TOTII, 1984).

Figure 1. The Unit Groundwater Basin: geometry, boundary conditions, unit patterns of hydraulic head h and flow $\vec{q}$, and areas of the three basic groundwater flow regimes. *Source:* Toth (1984).

groundwater moves beneath the site, water quality can change through time; the soil or sediment chemical quality generally will not change as rapidly. However, water moving vertically through the vadose zone from surface contaminated sites may significantly affect and change groundwater on arrival.

Changes in chemical analytical techniques and regulatory criteria affect the investigation since standards may change during the course of study. Analytical techniques have become more refined over the last ten years; resolution to parts per million and parts per billion is now possible. Thus, a constituent that was not detected at the parts per million range may now be detected and be of concern at the parts per billion range. As toxicological studies determine health risks at certain concentrations, these concentrations become critical to the investigation since they may or may not change the regulated standard. Regulations may change and their interpretation could affect the course of the investigation. Consequently, when a health risk is determined, it is likely that the regulation interpretation will become more conservative. If the regulated concentration of the contaminant is lowered, the cost and time of the investigation and cleanup will probably increase.

INORGANIC COMPOSITION AND QUALITY

Inorganic groundwater chemistry deals with the physical and chemical factors that govern groundwater movement. For detailed discussions of inorganic groundwater chemistry, refer to Hem (1970). Usually four criteria are used in a preliminary groundwater quality evaluation for almost all studies: specific electrical conductance, pH (acidity/alkalinity), total dissolved solids, and hardness (Table 1) (Heath, 1982). These measurements are typically taken each time the monitoring well is sampled. These chemical parameters yield quick and inexpensive chemical information and allow the hydrogeologist to chemically classify the groundwater at the time of well sampling.

Electrical Conductivity

Electrical conductivity (EC) refers to the ability of a substance to conduct electrical current (Hem, 1970). The ability to transmit an electrical current depends on the concentration of charged, or ionic, species in the water. Hence, the measure of the conductance is used to

Table 1. Characteristics of Water That Affect Water Quality

Characteristic	Principal Cause	Significance	Remarks
Hardness	Calcium and magnesium dissolved in the water	Calcium and magnesium combine with soap to form an insoluble precipitate (curd) and thus hamper the formation of a lather. Hardness also affects the suitability of water for use in the textile and paper industries and certain others and in steam boilers and water heaters.	USGS classification of hardness (mg/L as $CaCO_2$): 0–60: Soft 61–120: Moderately hard 121–180: Hard More than 180: Very hard
pH (or hydrogen-ion activity)	Dissociation of water molecules and of acids and bases dissolved in water	The pH of water is a measure of its reactive characteristics. Low values of pH, particularly below pH 4, indicate a corrosive water that will tend to dissolve metals and other substances that it contacts. High values of pH, particularly above pH 8.5, indicate an alkaline water that, on heating, will tend to form scale. The pH significantly affects the treatment and use of water.	pH values: less than 7, water is acidic; value of 7, water is neutral; more than 7, water is basic

Specific electrical conductance	Substances that form ions when dissolved in water	Most substances dissolved in water dissociate into ions that can conduct an electrical current. Consequently, specific electrical conductance is a valuable indicator of the amount of material dissolved in water. The larger the conductance, the more mineralized the water.	Conductance values indicate the electrical conductivity, in micromhos, of 1 cm^3 of water at a temperature of 25°C.
Total dissolved solids	Mineral substances dissolved in water	Total dissolved solids is a measure of the total amount of minerals dissolved in water and is, therefore, a very useful parameter in the evaluation of water quality. Water containing less than 500 mg/L is preferred for domestic use and for many industrial processes.	USGS classification of water based on dissolved solids (mg/L): Less than 1,000: Fresh 1,000–3,000: Slightly saline 3,000–10,000: Moderately saline 10,000–35,000: Very saline More than 35,000: Briny

Source: Heath (1982).

approximate the total concentration of ionic species present. Measurement of electrical conductance is usually in micromhos per centimeter (μmho/cm) or siemens per centimeter (s/cm). Usually the measurements are standardized to 25°C.

Many influences may cause EC measurements to change, and thus it is only a gross estimator of dissolved salt load or contamination. Field EC measurement equipment available today commonly utilizes self-calibrating adjustments or is calibrated to known standards when it is used.

pH

pH is the measure of the alkalinity and acidity of the groundwater, or hydrogen concentration, on a logarithmically calculated scale where 1 is the most acidic and 14 is the most basic (Gymer, 1973). pH has considerable influence on the water geochemistry because it affects ionic strength, oxidation-reduction, organic carbon content, and the mobility of metallic ions. This measurement is considered to be more accurate in the field since pH may change due to temperature changes, carbon dioxide or other gases escaping, or gases entering (API, 1983). For example, when sampling a stable reading of pH during water extraction, typically it is accepted as the true pH of aquifer water and is noted prior to the sample collection.

Oxidation-Reduction

Oxidation-reduction, also known as the redox potential (Eh), is the measure of the relative intensity of oxidizing or reducing conditions in solutions provided by the Nerst equation (Hem, 1985). When the pH is known, the stability of minerals in water can be determined. Measurements of dissolved oxygen may indicate whether or not groundwater has an oxidizing condition (API, 1983).

Total Dissolved Solids

Total dissolved solids comprise dissociated and undissociated substances in the water (Matthess, 1982). The value is commonly determined by evaporating a water sample to dryness, although the residue is slightly different from the solution due to minor losses and precipitation. This is a common indicator of overall quality.

REGULATIONS ESTABLISHING DRINKING WATER QUALITY CRITERIA

Standards have been set forth by the federal and state governments for the minimum drinking water quality for human consumption. The federal Office of Drinking Water has established recommended maximum contaminant levels (RMCLs) for contaminants in water. The RMCLs are health-based standards derived from toxicological data. RMCLs are health-related goals and are not enforceable drinking water standards. However, the federal Primary Drinking Water Standards has established maximum contaminant levels (MCLs) which are federally enforced. The MCLs are set as close as possible to the RMCLs, after taking into account technology available and cost to achieve or meet the drinking water standard. Individual states have adopted either the federal criteria or a modification of it as their water quality standard. These standards cover only a few of the potential contaminants that may affect water supplies (Table 2).

Additional water quality information is available from federal and state government sources regarding contaminants not covered under the aforementioned standards. Health advisories have been issued on certain chemicals that are derived from National Academy of Sciences and EPA information. The EPA has established the national ambient water quality criteria (NAWQC) under the authority of the Clean Water Act of 1974. NAWQC are not mandatory standards, but states can adopt them as enforceable standards to protect beneficial uses of water bodies.

NATURAL "CONTAMINANTS"

Although groundwater can be drinkable directly from the subsurface, it is not of drinkable quality everywhere, and in fact may need some kind of treatment in most areas today prior to distribution for potable use. Although the water may be usable, if it exceeds the current state or federal regulated chemical constituents, then it can be considered "contaminated" and unfit for human consumption. The influence of regional geology may have profound geochemical effects upon water quality. Since the observation of inorganic contaminants in water (and soil) may initiate a regulatory action, identifying naturally occurring contaminant sources is very important. For example, the Santa Clara Valley in California is adjacent to a large body of

Table 2.   National Interim Drinking Water Standards

Maximum contaminant levels for inorganic chemicals	
Contaminant	**Milligrams per liter (micrograms per liter in parentheses)**
Arsenic	0.05 (50)
Barium	1. (1000)
Cadmium	0.010 (10)
Chromium	0.05 (50)
Fluoride	2.2
Lead	0.05 (50)
Mercury	0.002 (2)
Nitrate (as N)	10.
Selenium	0.01 (10)
Silver	0.05 (50)
Standard	**Milligrams per liter (micrograms per liter in parentheses) except as noted**
Chloride	250
Color	15 units
Copper	1.0 (1000)
Corrosivity	noncorrosive
Foaming agents	0.5
MBAS (methylene-blue active substances)	
Hydrogen sulfide	not detectable
Iron	0.3
Manganese	0.05 (50)
Odor	3 (threshold no.)
Sulfate	250
Total residue	500
Zinc	5 (5000)

Maximum contaminant levels for organic chemicals	
Contaminant	**Milligrams per liter**
1. Chlorinated hydrocarbons:	0.0002
Endrin (1,2,3,4,10, 10-hexachloro-6,7-expoxy-	

Maximum contaminant levels for radium–226, radium–228, and gross alpha particle radioactivity	
1. Combined radium–226 and radium–228	5 pCi/L
2. Gross alpha particle activity (including radium–226 but excluding radon and uranium)	15 pCi/L

Radionuclide	**Critical Oxygen**	**pCi per liter**
Tritium	Total body	20,000
Strontium–90	Bone marrow	8

mercury ore–bearing rock. Over time, of course, the rock will erode, and the mercury may be released as sediment transported to the valley and, ultimately, may enter the groundwater via solution. If this metal is revealed as present in chemical analysis, then the water appears contaminated if present above the regulated standard. Since mercury is used in local manufacturing processes, is the observed source natural or anthropogenic? The solution may be simple if industrial development has not occurred upgradient; however, if urban sites are sold for redevelopment, a check for potential contamination could cause problems unless evidence is available to show it was not human-derived.

ORGANIC CHEMICAL CONTAMINANTS

The quality of groundwater, in terms of "organic quality" constituents, for this discussion relates to the presence of anthropogenic organic contaminants. Groundwater may contain some natural organic compounds. Organic carbon and humic and fulvic acids are naturally present in groundwater and arise from organic decomposition and other inorganic processes (National Water Well Association, 1986). Organic carbon concentrations are generally low in groundwater; due to long residence times the carbon oxidizes to carbon dioxide, contributes to alkalinity, or may recombine to form methane and be adsorbed onto aquifer material. For the purpose of the following discussion however, we will assume that groundwater has only natural inorganic components, and that organic materials found in groundwater have arisen from some anthropogenic contaminant source.

Information Sources for Organic Contaminants

The growth of the chemical industry has involved both refined natural petroleum compounds and synthetic chemicals and materials, and waste by-products of those processes. Consequently, there are literally thousands of compounds that may be released into the environment and contaminate soil and groundwater. Organic contaminants may have resulted from long-ceased operations, such as coal tar plants. Polychlorinated biphenyls (PCBs) have been used commonly since about the turn of the century and may be found almost anywhere through use of solvents, pesticides, chemical sealers, explosives, or rocket fuels. Organic contaminant history from processing or manufacturing may be complex and extend far into the past. EPA has prepared a list of the contaminants of most concern, commonly referred to as the Appendix IX List or the Safe Drinking Water Act (Table 3). Typically, a history of the use of chemicals for the site is a first investigation step since the ownership and chemical use may vary widely over time.

Information on subsurface fate, transport, and behavior in the groundwater environment is limited since only recently has organic contamination been recognized as a threat to environmental quality. Research concerning fate and transport models is currently ongoing. To date, few contaminants (motor fuel hydrocarbons and trichloroethylene solvents) have been studied. Given the complexity and plethora of industrial chemicals, the services of a chemical engineer and chemist familiar with the chemicals, their properties, and analytical techniques are required. Information on movement, material decomposition, toxicity, and aquifer adsorption is limited, making the design of fate and transport models difficult.

Types of Organic Contaminants

Obviously, almost any chemical could become a potential contaminant under the right circumstances, and the reader is referred to chemical indexes and dictionaries regarding specific compounds (Sax and Lewis, 1987; Montgomery and Welkom, 1989). The following contaminants are listed by groupings of EPA laboratory analytical methods taken from the EPA's *Methods of Analyses for Water and Wastewater* (1986b). These groups bring together contaminants with similar chemical properties and possibly similar industrial use, allowing the hydrogeologist to get a handle on the type of organic compounds present and attempt to predict their movement and behavior in groundwater. An example of a gas chromatography/mass spectrometry chromatogram identifying a volatile organic chemical "group" is presented in Figure 2. The following list is not meant to be comprehensive, but rather a list of the general chemical groups that often concern regulating agencies.

1. *Volatile organic compounds:* These typically are compounds with low vapor pressure and may include solvents, fuels with an aromatic chemistry (benzene ring type). These may include halogenated compounds (containing chlorine, fluorine, or bromine), such as chlorinated solvents and materials (trichloroethylene, methylene

Table 3.   Contaminants Regulated under Safe Drinking Water Act, 1986 Amendments

Volatile Organic Chemicals	Silver	Toluene*
Trichloroethylene*	Fluoride*	Adipates
Tetrachloroethylene	Aluminum	2,3,7,8-TCDD (Dioxin)
Carbon tetrachloride*	Antimony	1,1,2-Trichloroethane
1,1,1-Trichloroethane*	Molybdenum	Vydate
1,2-Dichloroethane*	Asbestos*	Simazine
Vinyl chloride*	Sulfate	Polynuclear aromatic
Methylene chloride	Copper*	hydrocarbons (PAHs)
Benzene*	Vanadium	Polychlorinated biphenyls
Chlorobenzene*	Sodium	(PCBs)
Dichlorobenzene(s)*	Nickel	Atrazine
Trichlorobenzene(s)*	Zinc	Phthalates
1,1-Dichloroethylene*	Thallium	Acrylamide*
trans-1,2-Dichloroethylene*	Beryllium	Dibromochloropropane
cis-1,2-Dichloroethylene*	Cyanide	(DBCP)*
		1,2-Dichloropropane*
Microbiology and Turbidity	*Organics*	Pentachlorophenol*
Total coliforms*	Endrin	Pichloram
Turbidity*	Lindane*	Dinoseb
Giardia lamblia *	Methoxychlor*	Ethylene dibromide*
Viruses*	Toxaphene*	Dibromomethane
Standard plate count	2,4-D*	Xylene*
Legionella	2,4,5-TP*	Hexachlorocyclopentadiene
	Aldicarb*	
Inorganics	Chlordane*	*Radionuclides*
Arsenic*	Dalapon	Radium 226 and 228
Barium*	Diquat	Beta particle and photon
Cadmium*	Endothall	radioactivity
Chromium*	Glyphosphate	Uranium
Lead*	Carbofuran*	Gross alpha particle activity
Mercury*	Alachlor*	Radon
Nitrate*	Epichlorohydrin*	
Selenium*		

*Included in EPA proposed and final rules published in the *Federal Register*, Nov. 13, 1985.

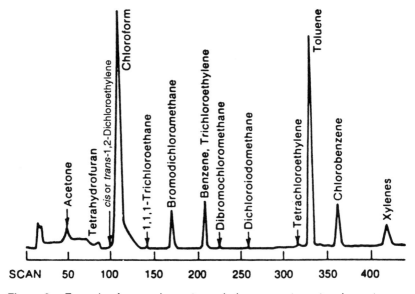

Figure 2.   Example of a gas chromotography/mass spectrometer chromatogram identifying volatile chemicals.

chloride, or chloroform). Depending on the compound, miscibility and solubility vary.

2. *Acid-base neutral compounds:* These may include polynuclear aromatics, ethers, esters, phenols, PCBs, plasticizers, and similar industrial compounds.

3. *Agricultural chemicals:* These include the vast array of pesticides, herbicides, nematocides, and related chemicals. These can be significant, given the quantity of chemicals used in modern agriculture that are applied directly to the ground. Some older compounds, such as DDT, are long-lived (months to years) but migrate conservatively; others are very short-lived (hours to days) but may be mobile in the groundwater environment.

4. *Trace elements:* These include selected trace elements (13 metals), asbestos, and cyanides.

5. *Alcohols and ketones:* These are common as cleaners and can tend to move rapidly due to their high solubility.

6. *Oils, greases, and petroleum products:* These cover a wide range of hydrocarbon materials used for fuels and lubricants. They tend to have a range of solubilities and are immiscible with water.

SELECTION OF ANALYTICAL LABORATORIES

The selection of the analytical laboratory used through the investigation is a critical decision. Once chosen, the laboratory will generate all the chemical data the consultant reviews and reports to the regulating agency. The accuracy, validity, and reproducibility of the laboratory results are vital to the success of any study. The laboratory should be able to perform all the analyses required by the investigation. Some contract laboratories may have specialty analyses, such as for pesticides, or may not do certain desired analyses. If the analyses and results are suspect, the data may be discounted and the concentration – or even presence – of the contaminant may be questionable. It also may create litigation problems if errors are encountered in methods, analysis, quality control, and chain-of-custody documentation.

Contract analytical laboratories usually perform analyses according to recognized standards. These include American Society for Testing and Materials (ASTM) and EPA standard methods and are reviewed by government and related professional societies and industrial research facilities. In some states, laboratories may be certified by state agencies; in other states, laboratories need only attest to the fact that they use approved EPA methods. If a government certification is required by the overseeing agency, then the laboratory must have the certificate prior to doing the analysis. It is also important to note that some compounds of interest to the investigation may not have "standard" methods of analysis. Some methods may be compound specific – for example, for certain pesticides or oils. Analytical methods may need to be modified depending on the type of contaminant and laboratory capability. Hence, it is advantageous to contact the laboratory director to discuss what work and analyses are desired prior to submitting samples for analysis.

Laboratory Selection Criteria

Over the past several years, a general procedure for laboratory selection has been evolved by EPA, which usually provides good criteria for laboratory selection. The following criteria summary is drawn from the EPA *Technical Enforcement Guidance Document* (1986a). The quality assurance and control procedures used by the laboratory must be thoroughly reviewed. Note that individual states may have guidance documents or criteria that have to be considered and referenced for a particular investigation. Additional guidance documents and material prepared by specific laboratories should be utilized in the investigation.

Individual laboratory quality control selection criteria may include analysis spike samples, duplicates, and specific standards to ensure the accuracy and precision of chemical analyses. The data must be reported accurately, with full laboratory documentation for valid results. Statistical and mathematical procedures for data reduction, instrument calibration, and matrix interferences should be reviewed to monitor performance. These procedures provide checks on cross-contamination but should not be utilized to alter or correct established analytical data. The individual analytical procedure or site sample matrix may cause site-specific deviations that must be taken into account when interpreting data.

Units of concentration should be consistent from report to report. Usually analysis results are reported in milligrams per kilogram, or liter, or parts per million. Depending on the laboratory, and the sample matrix (soil or water), analyses may range into the micrograms per liter or parts per billion range. Although this seems straightforward, many times confusion results if detection limits of the analytical instruments vary, or if the consultant requests a unit change for some reason. The data may be reported as "not detected," but this means only that it was not detected at the detection limits for that instrument for that analysis for that day. The units reported by the laboratory should always be those used and discussed in the investigation narrative and data reporting to avoid confusion.

LABORATORY DATA REPORTING AT THE DETECTION LIMIT

Since the water quality data will deal with contaminants in parts per million or parts per billion ranges, the accuracy of the analytical instrument during the analysis is critical. The analytical instrument may have varying accuracy on given days and the accuracy near the detection limit, or limit of accurate measurement, may have profound implications for the reported value. For example, if a contaminant is reported present at the same value as the detection limit, is it really there? Residual concentrations of a contaminant in an aquifer may create long sequences of very low concentration "hits." The laboratory report of contaminant "presence" may trigger a regulatory action based on that reported concentration. The laboratory should have checking procedures to ascertain whether it is present or if there are analytical problems. Extreme care should be used for any data reported at or near the detection limit of an analytical instrument, and discussion with the lab personnel should continue through all phases of analysis regarding chemical variations and data interpretation. When data are sent to the overseeing agency, the agency typically accepts the analytical information as correct, so the hydrogeologist should have confidence that the numbers are "real." Otherwise costly additional analyses or investigations may result when they may not be needed.

DATA MANAGEMENT

Both investigation and periodic monitoring programs generate huge quantities of information. A data management system is needed for each site so that the information is accessible and can be manipulated for needs of reporting and analysis for trends. This should be done at the initiation of site work so that the data are stored on a historical database, and the information can be retrieved when desired. If the project is long-lived, then well data is logged at quarterly intervals for chemical analyses as well as the quality control check analyses.

This is a practical problem in data manipulation. Computers and spreadsheet programs are useful for data storage. Computers and database spreadsheets are available, depending on the use and type of project. When linked with graphing programs, the data can be reduced to show detected contaminants and time trends. For example, concentration or separate phase product may be plotted against well water levels. Usually a data set showing one well with chemical data by sample date, laboratory sample number, and contaminant may be used in reports or required presentations. Data management methods should be thought out in the initial planning stage of the project if the anticipated database is large.

A CASE OF FALSE POSITIVES

A 1000-gal gasoline spill from a subsurface storage tank allowed gasoline to migrate through a very porous vadose zone and into an aquifer of sand and gravel. Transmissivity values for the aquifer range from 100,000 to 1,000,000 gpd/ft. A regional groundwater pumping field about 1200 ft away pumped over 2,000,000 gpd for potable water consumption. The bulk of the spill was cleaned, but some product entered groundwater and monitoring wells were installed to ascertain benzene migration to the pumping wells. A simple benzene linear transport model predicted a time to observe whether the plume was, in fact, moving to the pumping wells.

During three months of weekly monitoring contaminants were not detected. The contaminant should not have passed the monitoring wells based on the assumed groundwater velocity. The plume was not observed in the monitoring well array. Then one sample showed a one-time occurrence of very low concentrations of xylene. According to the benzene transport model assumptions, the xylene would indicate the trailing part of the plume. Benzene, which should have preceded the plume, was never detected in any sample at any time. Given the one "hot" xylene sample and potential threat to a large water supply, the regulating agency was ready to order immediate extraction well installation and counterpumping. The cost for one extraction well and initial emergency pumping could exceed $100,000; was extraction really needed based upon the laboratory data?

The well was resampled and the results were similar, but near detection limit. Sample blanks were taken for the sampling equipment and travel, and a duplicate sample was sent to another laboratory. The second laboratory showed that contaminants were not detected in either blank or second sample. The appearance from the data was that the first laboratory had erred—either by contaminating the sample internally, or by not properly cleaning their instrument and inducing contamination in the sample. Now it is up to the judgment of the professional—if additional xylene hits are observed, a plume moving toward the wells would be assumed. The immediate need for extraction well installation did not seem necessary given the previous not detected data. The model-predicted plume would reveal benzene first and xylene last. Additional sampling revealed all the fuel constituents for which the program monitored were not detected. One data point does not indicate a trend, and additional monitoring must now be made to confirm this. The low level of xylene observed was attributed to possible lab error, and xylene presence was not observed again over several additional months. The observed data were assumed to be a one-time occurrence. Contaminant plume location needs to be verified by continually monitoring presence in wells. By narrowing the problem to possible laboratory error and lack of groundwater contaminant trends, a costly remediation that was not needed was avoided.

The point is that often limited data must be evaluated and reported. One piece of data may initiate a regulatory action. Errors may occur at any point in sampling and analysis, and a resample and laboratory check is necessary. The consultant's approach to the problem may need adjustment also. Laboratories may make mistakes or report data at very low concentrations that, based on a one-time occurrence, are extremely difficult to substantiate. Of course, analytical laboratories should not become scapegoats just because the data do not make sense to the hydrogeologist. Analytical data have to be taken in the context of the spill problem, sampling, analysis, aquifer conditions, and all observed data trends. Taken together, these may signal presence of a contaminant plume, or a "ghost." Data extrapolation, given limited information, is always a difficult problem for the consultant to resolve.

CHAPTER 7

Contaminant Pathways, Subsurface Investigation, and Monitoring Approach

INTRODUCTION

When a contaminant is introduced at the ground surface, it must migrate through the unsaturated zone toward the aquifer. Hence, contaminants will move through soil, sediment, fractured rock, human-made conduits, or other pathways on their way to the saturated zone. Investigations need to ascertain the horizontal and vertical extent of contaminants in both unsaturated and saturated zones. Additionally, the existing soil and groundwater quality must be quantified for background comparisons. It is important to establish preexisting contamination or natural background that may exceed a regulated standard.

TYPES AND SOURCES OF CONTAMINANTS

Contaminants are "unnatural" substances introduced in the subsurface that degrade natural groundwater quality. Obviously, a complete list of contaminants is beyond this discussion; however, the following list of contaminant groups is presented. Classifying contaminants into groups may assist the investigator since industrial or other processes may utilize certain materials that give rise to working contaminant "suites" (for example, solvents and certain heavy metals for microchip manufacture, or hydrocarbons at a refinery or service station). Such contaminant groups may include inorganic heavy metals and organic materials, including hydrocarbon fuels, greases and oils, industrial solvents and chemicals, herbicides and pesticides, coolants, explosives, and agricultural chemicals and wastes (EPA, 1986a, 1987b).

The sources of these materials are widespread and numerous.

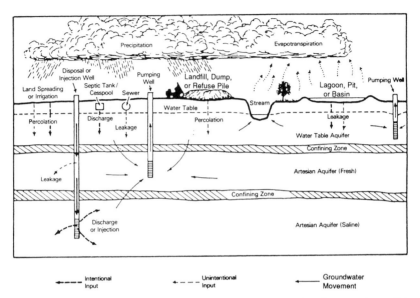

Figure 1. Conceptual input of contaminants into the subsurface and possible pathways of migration. After Geraghty and Miller (1985) in EPA (1987).

Again, a classification grouping may tie together the types of contaminants of interest for the investigation (Geraghty and Miller, 1985):

municipal: sewers, sanitary landfills, disposal wells, military bases

industrial: manufacture, subsurface tanks, pipelines, mines, oil fields

agricultural: fertilizers, pesticides, wastes, irrigation return flow

other: transport spills, material stockpiles, septic tanks, testing labs

Once these materials migrate from a source, they must move along some route, or pathway, into soil or rock and toward groundwater (see Figure 1).

CONTAMINANT PROPERTIES AFFECTING TRANSPORT

The chemistry of the contaminants will affect their transport and fate. Moore and Ramamoothy (1984) divide these properties into two groups:

1. physicochemical properties, such as solubility, vapor pressure, partition coëfficient, sorption/desorption, and volatilization

2. chemical transformations, such as oxidation-reduction behavior, hydrolysis, halogenation/dehalogenation, and photochemical breakdown

These processes will be briefly discussed below, and this following review borrows heavily from the work of Moore and Ramamoothy (1984).

Physicochemical Processes

Solubility is the degree and ease with which the chemical or compound will dissolve in water. Solubility determines the concentration present in water and whether the contaminant will interact with other chemicals. The precise determination of solubility remains elusive for many contaminant compounds, and some of the aqueous solubility values are only estimates. Many environmentally sensitive compounds have very low water solubilities.

Vapor pressure can be considered related to the solubility of the compound in air from the liquid phase. Chemical data charts and reference books are available that contain vapor pressure data.

The partition coefficient is a measure of the distribution of a given compound in two phases and is expressed as a concentration ratio, assuming simple dissolution. In reality, the situation could be more complex as a result of molecular changes.

Sorption and desorption, as stated by Moore and Ramamoothy (1984), means that the more hydrophobic the organic compound is, the more likely it is to be sorbed to the sediment. The solubility of an organic compound depends on the characteristics of the compound and sorbent geologic matrix. Sorbent characteristics of the geologic matrix include surface area, nature of charge, charge density, presence of hydrophobic areas, presence of organic matter, and strength of sorption. Sorption can be expressed by

$$C_s = K_p C_w^{1/n}$$

where C_s = the concentration of the organic compound in solid phase

C_w = the concentration of the organic compound in water phase

K_p = the partition coefficient for sorption

$1/n$ = an exponential factor

Volatilization is the transport of a compound from the liquid to the vapor phase and is an important pathway for chemicals with high vapor pressure or low solubilities.

Chemical Transformations

Oxidation and reduction reactions (redox) involve the liberation of electrons (oxidation) and reactions that consume electrons (reduction). Many organic compounds can either accept or donate electrons. This is environmentally significant since the oxidized or reduced forms of organic compounds may have totally different physical and/or chemical properties.

Hydrolysis involves the reaction of hydrogen, hydroxyl radicals, or water molecules with the organic compound, depending on the pH and polarity of the reaction site on the molecule.

Halogenation/dehalogenation of organic compounds occurs mostly under synthetic conditions or in drastic environments. Moore and Ramamoothy (1984) state that mild chlorination reactions are possible in natural waters with effluents that contain residual chlorine. Dehalogenation may occur under varying reactions of hydrolysis or disproportionation.

Photochemical breakdown processes involve structural changes in a molecule induced by radiation in the near ultraviolet–visible light range. The structure of an organic compound generally determines whether or not a photochemical reaction is possible.

CONTAMINANT PATHWAYS THROUGH THE VADOSE ZONE

Releases from buried tanks, pipelines, building basements, or any subsurface source can promote the migration of contaminants since these locations are below the surface, where direct observation is not possible. These releases place material directly in the subsurface vadose zone and can promote vertical movement, since they are below ground surface, and decrease transport time if groundwater is shallow. Many recent laws and regulations have addressed the need to monitor potential subsurface release points. A prolonged leak situation may occur for years before discovery, at which point a widespread problem may have already evolved.

Contaminant movement through the vadose zone will tend to follow fluid migration according to the movement processes discussed

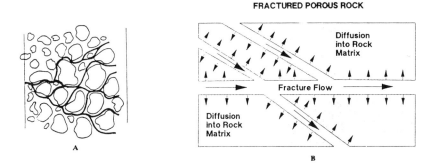

Figure 2. Flow paths that cause hydrodynamic dispersion in the subsurface: A. porous media (Fetter, 1988); B. fractured rock (EPA, 1989).

previously. If cracks or fractures (macropores) exist, the contaminant may flow directly into the crack and move downward. This can also happen if the contaminant material is a solid, such as a powdered material moved by the wind, and settles on a crack-containing surface. This is a very important fluid pathway if the site is covered by clayey soils, which readily desiccate, allowing cracks to form, often feet deep. Thus, just because the site is covered with a clay (implying low permeability cover retarding movement) does not preclude a possible rapid fluid or contaminant infiltration movement (see Figures 2 and 3).

Movement through the subsurface may be enhanced by biologic structures and voids in soil or sediment. The action of burrowing flora and fauna may leave extensive networks of open holes and voids, as well as bioturbating the sediment. Although these voids may only be tenths of an inch in diameter, they can transmit fluid in both horizontal and vertical directions. Additionally, these voids may be open to depths of tens or even hundreds of feet in thick alluvial fills, so assuming that voids shrink or close with increasing depth may not always be true. Finally, voids may be filled in with different types of sediment, such as sand in clay, thus increasing hydraulic conductivity in an otherwise lower-permeability stratum. Morrison (1989) has reviewed a pesticide transport model to ascertain how the presence of open worm tubes enhances macropore flow. These tubes allow rapid movement due to "instabilities" where preferential flow may occur as a vertical fingering. The flow in the "fingers" tends to coalesce with depth, so fluid movement is more rapid in vertical directions, with a slow horizontal wetting from the fingers. Morrison (1989) also points

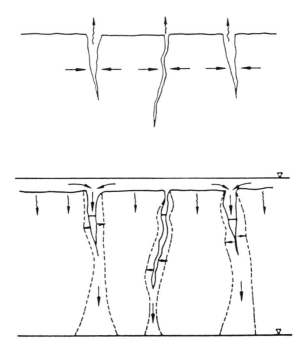

Figure 3. Fluid entry and penetration through surface mudcracks, causing macropore flow and rapid vertical movement. *Source:* EPA (1985).

out that although this phenomenon is somewhat texture dependent, it is related to the pore size, number of irregularities (openings), and initial moisture contents. The potential to skew vadose model flow assumptions in computer or mathematic models is obvious.

Soil vapor from volatile contaminants moves in the subsurface away from the source (Schwille, 1988). The movement may occur though natural deposits or through human-made conduits such as utility trenching, pipelines, into and under structures and vents to the ground surface, or basements. The problems of migrating potentially explosive vapors or formation of unbreathable atmospheres have been known for some time. However, using soil vapor for subsurface investigation is a relatively new technique and is widely used as a reconnaissance tool. Basically, soil vapor surveys are done by placing probes 5 to 15 ft into the vadose zone and extracting vapor samples. The vapor is then analyzed in a portable analyzer (organic vapor or gas chromatography), and a concentration contour map of vapor occurrence is plotted on the site plan. The vapor survey will generally "overprint" the groundwater contaminant plume. The isoconcentration contours are used to infer the presence of the contaminant in soil or groundwater, and possible exploratory boring locations are chosen from the vapor survey data. This can save money and time by placing exploratory borings closer to suspected sources to define unsaturated and saturated plume extent.

When contaminants move through the subsurface, they often leave a residual portion in the available porosity. Presence of residuals are not easily found or removed and may constitute a long-term source of contamination. Wilson et al. (1988) have conducted column experiments examining the affect of porosity and pore throat morphology on organic liquids moving through porous media. Their results indicate that residual liquid saturations are influenced by liquid properties that exceed a critical capillary force, and that residual saturations may result from the presence of air (as a nonwetting phase) and larger buoyancy forces and smaller capillary forces. In a sandy medium, average residual organic liquid saturations were 29%, and in the dry range of vadose zone was 9%. If suction is high or contaminants are trapped by water droplets, the contaminant can become immobilized, as shown in other glass bead experiments (see pages 88 and 89 in Schwille, 1988).

CONTAMINANT MOVEMENT WITHIN THE AQUIFER (SATURATED FLOW)

Contaminant migration in the aquifer depends upon the properties of the contaminant, aquifer geology, and groundwater velocity. The following discussion will focus primarily upon stratigraphic approaches in either alluvial or sedimentary rock geologic terrains. The three general migration pathways commonly used in investigating conceptual aquifer models are the floaters (immiscible contaminants), mixers (contaminants with uniform dissolution and movement in the aquifer), and sinkers (contaminants that move vertically due to density makeup) (see Figure 4). These general conceptual models are useful in drawing simplified models; however, as with all real-world situations, the subsurface reality is more complex. The chemical composition, molecular weight, solubility, and viscosity will also influence contaminant movement. For example, hydrocarbon fuels are commonly called floaters, although several common industrial solvents may fit the floater model. Conversely, the sinkers are

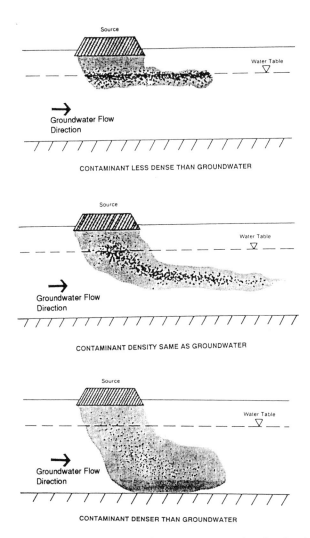

Figure 4. Three generalized contaminant transport paths showing influence of contaminant density.

conceived of as industrial chemicals and solvents, whereas all chemicals classified this way may have properties that allow some mixing and sinking. Generalizations are useful, but often mislead, so the site-specific environments and contaminant chemical characteristics must be taken into account.

CONTAMINANT PLUME CONFIGURATION AND MOVEMENT

Fetter (1988) describes the basic groundwater transporting processes for contaminant plume solutes as advection and diffusion. *Advection* is the process by which groundwater in motion carries dissolved solutes. *Diffusion* is the process by which both ionic and molecular species dissolved in water move from areas of higher concentration to areas of lower concentration.

Advection is the rate of flowing water as determined by Darcy's law:

$$V_x = \frac{Ki}{n}$$

where V_x = average linear velocity
         K = hydraulic conductivity
         n = effective porosity
         i = hydraulic gradient

Fetter (1988) describes the diffusion of a solute through water as described by Fick's laws and the flux of a solute under steady-state conditions as

$$F = -D \frac{dC}{dx}$$

where F = mass flux of solute per unit area per unit time
         D = diffusion coefficient (area/time)
         C = solute concentration (mass/volume)
   dC/dx = concentration gradient (mass/volume/distance)

As the plume moves, processes of dispersion and retardation influence its size and shape. Mechanical dispersion occurs as the contaminated fluid flows through and mixes with noncontaminated background water. Dispersion is caused by the differing fluid velocities within the pores and pathways taken by the fluid (see Figure 5) (Fetter, 1988). This mechanical dispersion on a microscopic scale is due to a microscale deviation from the average groundwater velocity; Anderson (1984) reports that several investigators maintain that in order to apply advection-dispersion equations, dispersivity must be defined in terms of the statistical properties of hydraulic conductivity for an aquifer of a given size.

Molecular diffusion occurs as species move from higher to lower

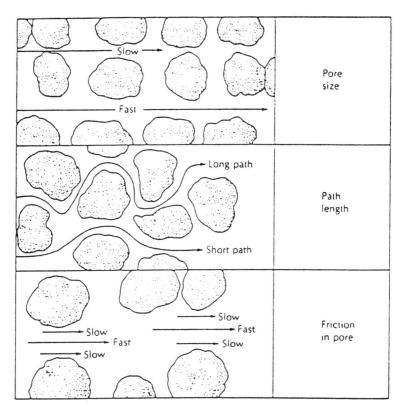

Figure 5. Factors causing longitudinal dispersion. *Source:* Fetter (1988).

concentrations on the microscopic scale (Anderson, 1984). Cherry and Gilliam (1984) state that molecular diffusion is important in contaminant transport in fine-grained deposits. Diffusion may also become important in contaminant transport in heterogeneous deposits and low flow velocity, allowing diffusion from higher to less permeable strata (Cherry and Gilliam, 1984). The plume shows divergence in the longitudinal and lateral groundwater flow lines. Fetter (1988) states that mechanical and molecular dispersion cannot be separated in groundwater flow regimes. Consequently, a factor termed the *coefficient of hydrodynamic dispersion* is used to take into account mechanical mixing and diffusion (see Figures 4–6).

The length and width of the plume will tend to move fastest in the downgradient direction, spreading laterally as a result of the aforementioned factors and aquifer texture (EPA, 1985b; Fetter, 1988). If

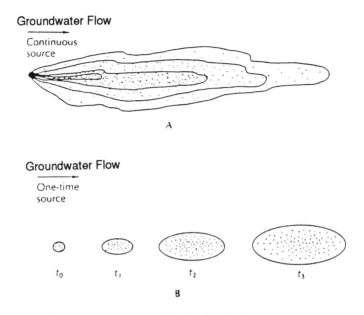

Figure 6. Contaminant plumes: A. developed from a continuous source; B. movement as a one-time "slug." *Source:* Fetter (1988).

flow velocity is low, the plume tends to be somewhat less elongated than at higher velocities. If the hydraulic conductivity is low, the plume moves slowly and stays relatively compact. Higher hydraulic conductivities may result in more rapid movement and a longer and narrower plume.

The plume configuration also depends on the type and chemistry of the contaminant. If it is immiscible, or a (nonaqueous-phase liquid) "floater," then a separate phase and dissolved phase will occur near the upper portion of the aquifer (such as gasoline or oil floating on the capillary fringe and groundwater surface). For example, separate-phase product will migrate to the capillary fringe until the weight of the product overcomes the capillary pressure and the product collects and flows on the water surface. When sufficient product collects on the groundwater surface, the surface is displaced downward (see Figures 7a, 7b, and 8) (Hughes, Sullivan, and Zinner, 1988). The separate-phase product displaces groundwater entering deeper porosity as groundwater depth fluctuations. A portion of the contaminant will dissolve in groundwater and migrate by advection, usually the benzene, toluene, ethylbenzene, and xylene (BTEX) constituents of

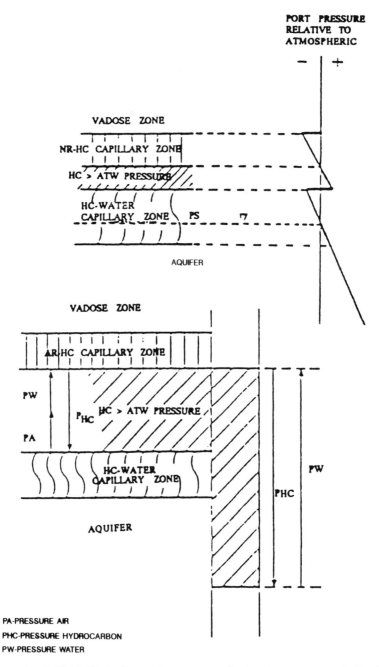

Figure 7. A. Fluid distribution and pressure profile in a homogeneous isotropic formation. B. Balance of forces in the formation and in the well.

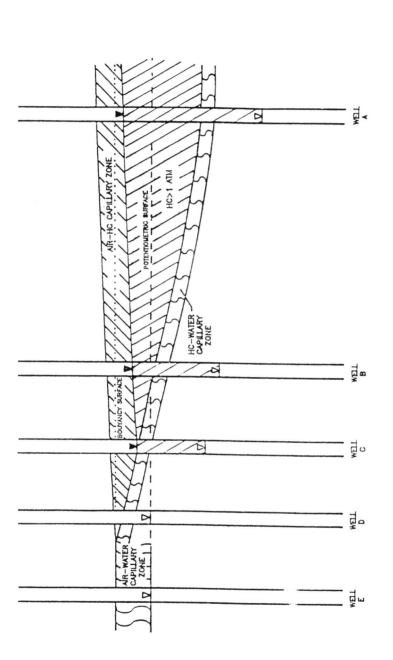

Figure 8. Well to formation hydrocarbon thickness ratios in a cross section of a hydrocarbon pool. Note relationship of observed product thickness in well versus the "actual" product occurring on the capillary fringe. Hence, apparent thickness of product observed in well can be highly misleading. *Source:* Hughes, Sullivan, and Zimmer (1988).

fuel (gasoline) and the other hydrocarbons contained in fuel (gasoline or diesel). The dissolved- and separate-phase products tend to be most concentrated in the upper portions of the aquifer.

Contaminant chemistry and solubilities may allow for mixing (or without some preferential partitioning) through the aquifer. This type of contaminant plume may become distributed into relatively uniform concentrations, once the plume has moved away from the source. A contaminant may have some preferential partitioning; however, if large quantities are released or it has been in the aquifer for a long period, concentrations may become somewhat evenly distributed.

A "sinker" (or dense nonaqueous-phase liquid) contaminant plume may show a concentration gradient through the aquifer, becoming more concentrated near the aquifer base. If a sufficient quantity of sinker enters the aquifer, a separate phase may collect at the aquifer base. These may be caused by solvents (volatile chlorinated hydrocarbons, CHC) that are denser that water. Model column experiments by Schwille (1988) indicate that CHCs will sink given enough CHC fluid pressure in unsaturated and saturated media. CHC penetration occurs after it has developed sufficient head to drive out water (in saturated media), and it is possible that CHCs will not effectively penetrate moist and heterogenous soils due to oversimplified assumptions of CHC density and viscosity properties (Schwille, 1988). Recent column experiments by Abdul, Gibson, and Rai (1990) were done to evaluate the flow of organic solvents through kaolin and bentonite. Their results indicate that aqueous solvents did not change the physical appearance of clays, and solutions moved through the clays at a constant hydraulic conductivity. However, hydrophobic solvents caused clays to shrink, forming networks of cracks and allowing fracture flow to occur through the clay.

Contaminants may partition, or separate, once dissolved into groundwater. For example, gasoline (a fuel that may contain tens of individual compounds) may partition into benzene, toluene, xylene, and other hydrocarbons; this type of partitioning is often observed at fuel spill sites (see Figures 9 and 10). CHCs such as trichloroethylene may similarly partition into dichloroethylene breakdown products that migrate at different velocities.

Fetter (1988) states that there are two broad classes of solutes: conservative and reactive. Conservative solutes do not react with the soil or groundwater and do not undergo decay (for example, chloride). Reactive substances can undergo chemical, biological, or radio-

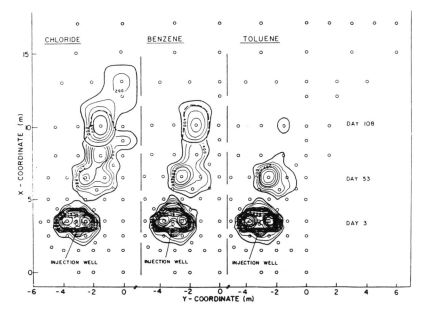

Figure 9. Injection experiment to observe effects of sorption, retardation, and biotransformation on organics (mass loss of benzene and toluene) compared with chloride. *Source*: Barker, Patrick, and Major, (1987).

active change (degradation), which tends to decrease the concentration of the solute. If a contaminant undergoes degradation, a single reaction or a series of degradation reactions may occur. For example,

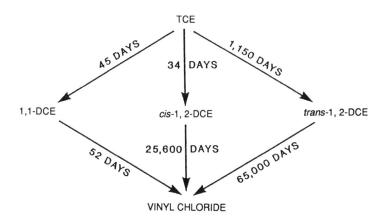

Figure 10. Possible degradation sequence of trichloroethylene in subsurface. *Source:* EPA (1987).

experiments by Barker et al. (1987) show that benzene, toluene, and xylene will migrate at different velocities and undergo biotransformation and degrade, given sufficient supplies of oxygen in the aquifer. Other degradation processes may include temperature, microbial processes, oxidation and reduction, instability of the compound in question, and reaction to form a new compound (such as chloroform or trihalomethanes). Further, the breakdown may proceed to methylene chloride, which is more "toxic" than the original trichloroethylene contaminant (see Figures 11 and 12).

EFFECTS OF AQUIFER STRATIGRAPHY ON CONTAMINANT MOVEMENT

The occurrence of sand and clay (higher or lower conductivity units) will profoundly affect movement in the aquifer. For example, thin clay beds may split, retard, or deflect flow, causing the plume to spread horizontally and vertically. Similarly, if the porosity is clogged by fine-grained matrix, the plume may move slowly, and move at lower flow rates than calculations may imply. Sand lenses may allow more rapid movement. Presence of clay may cause adsorption and contain a residual contaminant source. As groundwater rises and falls in the aquifer, the contaminant can be spread vertically and later be trapped by water refilling the available porosity, creating a widespread source throughout the aquifer.

Inferring a simple "layer cake" stratigraphy and utilizing overly simplistic homogenous conditions can be quite misleading. Conceptual flow paths must be based upon the actual site geology and hydrogeologic constraints. This relates directly to the accuracy and completeness of lithology logging collected during preliminary subsurface investigations. The chemistry of existing background and introduced materials must be factored into the conceptual geologic and groundwater flow model so that both field investigations and data analysis are meaningful (see Figures 13 and 14).

CONTAMINANT MOVEMENT IN FRACTURED ROCK

Contaminant movement investigations can be very complicated in geologic terrains containing fractured rock and sediments. The fluid movement through the fractures makes aquifer analysis and modeling

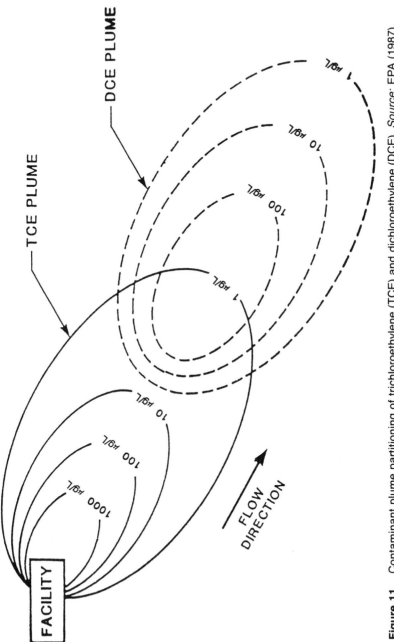

Figure 11. Contaminant plume partitioning of trichloroethylene (TCE) and dichloroethylene (DCE). *Source:* EPA (1987).

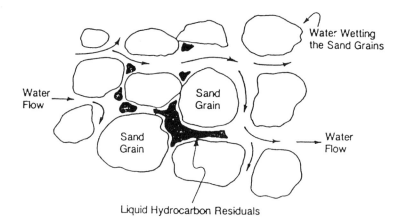

Liquid Hydrocarbon Residuals Trapped In
Water–Wet Sand Grains.

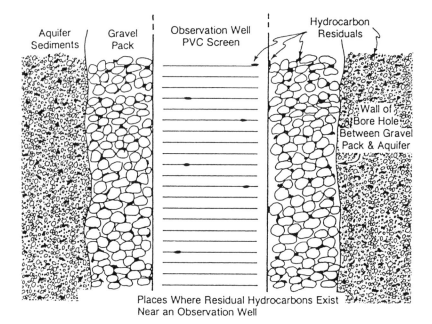

Places Where Residual Hydrocarbons Exist
Near an Observation Well

Figure 12.

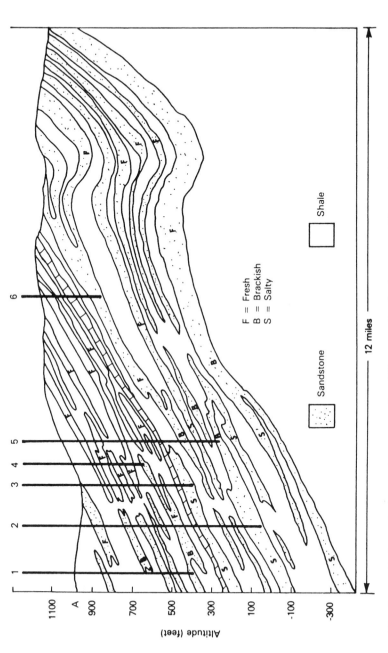

Figure 13. Conceptual monitoring and water quality differences in dipping strata. Since sedimentary formations may contain changeable strata, composition monitoring may reveal geochemical facies changing in those strata. Also, placing a hazardous waste facility on such a geology could intercept numerous different strata, some of which either enhance or retard fluid movement. *Source:* EPA (1987).

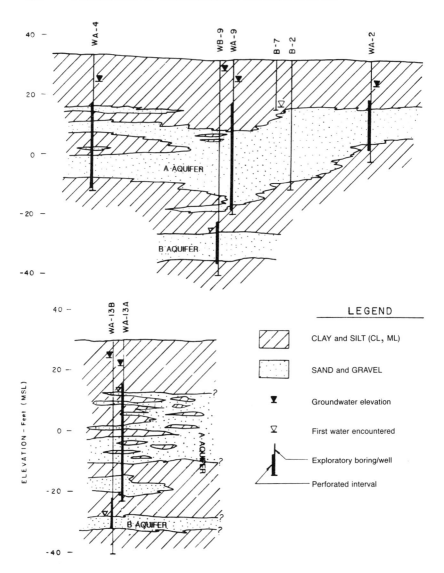

Figure 14. Cross sections beneath an industrial site reveal complex stratigraphy created by sediment deposition. This may enhance or retard contaminant flow, and makes location of the next underlying aquifer difficult. Several wells or borings may be needed to delineate stratigraphy in a small area.

contaminant movement very difficult. Schmelling and Ross (1989) have prepared a summary of contaminant movement in fractured media, from which the following is summarized. Fluid movement in most fractured rock systems is through fractures, joints, cracks, and shears, which may occur in sets or zones. Fractures may be open or filled in with mineralization. Shears or faulting movements may produce gouge or slickensides, which may form "skins" that can retard fluid movement. Groundwater movement will depend on the fracture density, interconnections, orientation, aperture width, and nature of the rock matrix. Igneous and metamorphic rock usually have low primary porosity and permeability, so fractures form the primary fluid pathway. Sedimentary rocks may have higher primary porosity and permeability, and fracture permeability is secondary; however, depending on cementation and induration, fractures may form the primary permeability. Schmelling and Ross (1989) state that rates of contaminant migration into or out of rock will depend on matrix permeability, presence of low-permeability fracture "skins," and the matrix diffusion coefficient of the contaminant.

CONTAMINANT MOVEMENT BETWEEN AQUIFERS THROUGH AQUITARDS

Aquitards that separate individual aquifers are impermeable in the conceptual sense only, and they transmit fluid, albeit very slowly. Toth (1984) shows that while hydraulic conductivities of clay or any low-permeability unit are low, flow will occur through these units. Additionally, the geochemical nature of the water will change due to the thermal and chemical interaction during flow. Groundwater geochemistry will show changing composition of water moving naturally when modeled in Toth's unit basin (Toth, 1984). Thus, natural chemical changes occur as water migrates through the aquifer and aquitard system. Natural leakage through aquicludes and aquitards allows groundwater to move, and it may or may not significantly change the geochemical character of aquifer water. Behnke (personal communication, 1989) reports that extreme pumping of aquifers may cause aquitards to compress, releasing water that may mix into and change aquifer water chemistry.

When contaminants enter the system, they may move as a natural chemical system would. However, movement can be enhanced since the materials can be introduced at any depth and in any geologic unit.

The most recognizable case is through a well that interconnects several aquifers, creating an opening through which groundwater may be transmitted vertically (see Figure 1). Contaminant movement can be accelerated by pumping wells, which draws the contaminant toward the well while mixing contaminated with noncontaminated water.

Wells may have been used for illegal disposal by dumping the unwanted material down the well. Since regional well records may not be accurate or complete, exact locations are unclear, and the driller's log may be incomplete, with aquifer contacts not well-defined. Construction details are often incomplete or missing. Often wells are abandoned with pumps and lines in the casing which rust and collapse, precluding future sampling. Finally, agricultural wells may tap several aquifers for maximum yield, followed by industrialization of the area. If the land is rapidly developed, wells may be buried and forgotten, leaving many points of aquifer interconnection, which will require considerable analysis and effort to relocate the wells. Older well casings can corrode, and the borehole may collapse but still allow water movement between different strata. Once relocated, the well must be properly cleaned and reconstructed, or abandoned and grouted shut.

NEEDS AND GOALS OF THE HYDROGEOLOGIC INVESTIGATION APPROACH

The ultimate goal of this type of hydrogeologic study is to ascertain the extent of the contamination so that an approach for remediation can be developed. The needs and goals of the investigation and the site conditions must be considered to effectively position exploratory borings and monitoring wells. It is also assumed that the regulators have reviewed the work plan (if required) and approved it prior to starting field work. Site-specific knowledge of aquifer stratigraphy, and contaminant type and suspected pathway will greatly influence well design and investigation procedures. The field rationale for well placement should then address the following:

define site geology and stratigraphy

define hydrogeology (water occurrence gradient and flow)

collect soil samples for vertical and horizontal vadose contaminants delineation

collect groundwater samples to define the plume extent

allow flexibility in data collection for overall site coverage

It is very important to remember that the hydrogeologist wants to gain the maximum information for the time and money spent in the field that is required to answer the regulatory questions and solve the client's problem. Groundwater quality changes as it moves through the contaminated area and influences downgradient quality. Otherwise clean sites may become contaminated from the upgradient plume sources. Since a property or political boundary may be crossed by the water, the problem grows in terms of responsible parties and their potential liabilities. The longer the plume moves and disperses, the larger the contaminated area becomes and the more comprehensive and complex the geologic investigation.

LOCATING THE CONTAMINANT PLUME

The Initial Approach

In order to attempt to locate the plume, the hydrogeologist will have to make some preliminary assumptions. Assuming the contaminant is present in the aquifer, the next assumption is that the groundwater chemistry is affected and moving at some average linear velocity. The hydrogeologist may review previous investigations' reports, soil vapor surveys, surmised contaminant properties, and any other available data. Ultimately, a subsurface investigation will be conducted to provide the geologic data and locate and install the permanent monitoring points. The money and time that will be committed to the subsurface study may vary depending on the level of effort, and typically the investigation is phased so that time and money are more effectively spent. Usually basic groundwater and contaminant chemistry information must be collected first.

Exploratory soil borings will be drilled and monitoring wells installed to collect the basic data. Some boreholes may be advanced to collect lithologic and chemical sampling information only. However, if groundwater gradient and flow direction are to be known, at least three monitoring wells are required. If this is the initial site investigation, then one well is placed in the surmised upgradient direction and two are placed in the suspected downgradient direction. These wells will allow a measurement of gradient to be determined, and additional wells are installed to detail the hydrogeologic and chemical data requirements. Groundwater contamination investigations are usually long affairs, and it is probable that numerous wells will be installed to define and monitor the plume (crossgradient and downgradient). Complications may arise if wells cannot be positioned in the most desirable locations due to access or legal problems, resulting in long distance extrapolation of subsurface data. Finally, the geology may or may not be easy to resolve or interpret, which requires additional subsurface work.

Ultimately the number of wells and level of effort required to define both unsaturated zone and groundwater contamination largely depends upon what is revealed by the initial field studies. The project will probably require several field phases, and the data collected from each step must be targeted for contaminant delineation and cleanup. Although the effort may take the appearance of a "research project," it is, in fact, an investigation to solve the client's problem — no more, no less. The amount of scientific effort and analysis must be sufficient to get the required answers.

Data Analysis and the Report

Once the field work is done, the information must be organized and analyzed to determine whether or not additional work is needed to complete the report and begin remediation. Some type of formal reporting document will be needed at this point. Each formal letter or report becomes a legal document, which the responsible professional must be able to defend technically and legally. All work must, of course, also be of the highest ethical standards. There are times when the investigation is preliminary or of a reconnaissance level. If this is the case, the report should clearly state the limited detail of scope. This may become an issue with implications regarding completeness or omission, unless the report actually states the limited, or comprehensive, scope of work. An excellent review of the approach to subsurface investigations and some of their legal implications is published by the Association of Engineering Geologists (1981).

Usually local, state, and federal regulatory guidance documents will set the requirement for a written report. It has been the authors' experience that the following sections are typically included in order to present the subsurface information in a format for delineating site geology and contaminant extent. These parts would comprise the minimum content, with other sections as needed to explain or address all the investigation issues. Hence, the following list of sections should be included:

Site plans or maps show monitoring well locations, site surface geology and geography, equipment (tanks, etc.), and any other required information. Several maps may be needed to represent the information. The maps should show the necessary information and cite sources of previously obtained information, especially if maps, tables, or cross sections are adapted from a published or unpublished report. Maps derived from preexisting sources should contain the reference on the map proper.

Cross sections show the site stratigraphy, aquifers, boring locations, well details, and other relevant geologic and hydrogeologic information. The elevation should always be set to mean sea level. Chemical data are often included, especially from vadose zone chemical sampling. All vertical and horizontal changes in stratigraphy and contaminant changes should be shown using as many sections as may be required.

Groundwater elevation contour maps display the groundwater contours and flow lines for the day the data were collected (typically the date on which the wells were sampled). The elevation should always be referenced to mean sea level.

Groundwater contamination contour maps show the distribution of the contaminants in water in relation to the site boundaries and contaminant source. Concentrations are often shown in logarithmic concentration contours since the values are commonly parts per million or billion. The concentration contours can also represent data from cross sections where appropriate.

Data tables present the large quantity of chemical and other data generated by the investigation. A table format makes the data readable and concisely presented. Presentation by well and date of sampling is usual and can be referenced to the narrative, figures, and graphs.

Report narrative will include all methods, procedures, and results of the study, along with the backup documents. The basic sections will include the geology, hydrogeology, extent of pollutants, and interpreted condition of the affected area. This may or may not have input from other consultants and clients, and even lawyers. The language used and the words chosen must convey what the site study has revealed clearly and without ambiguity. Accepted geologic and hydrogeologic terms should be used to describe the subsurface. Where the opinion of the responsible professional is given, it should be stated as such. The definitions of words may cause problems in the future, and so the choice of words may become a point of debate (for example, *observe* rather than *examine,* or *suggest* rather than *indicate*). Al-

though this may seem to hedge about the issue, wording, unfortu-
nately, is very important when considering legal documents. The wording should not confuse or compromise the basic scientific integrity of
the work or its conclusions. The report may also have recommendations sections; however, the data and work must support both conclusions and recommendations in all cases. Finally, all citations to other
work, especially other consultants' work, and appropriate legal regulations should be included both in the text and in separate citations or
reference lists.

The author should consider whether all the information makes sense.
Do the exploratory boring logs substantiate the correlation of soil/
alluvium/rock units? Were a sufficient number of soil and groundwater samples analyzed? Do groundwater maps and elevation contours agree with regional flow concepts? If not, why? Are chemical
results in agreement with observed trends or expected values, given
the hydrogeology and contaminant type? Do the monitoring wells
and chemical data accurately locate the plume? Where are the data
gaps? Which issues are resolved, and which are not? This review and
questioning should take place throughout the investigation and report
drafting. Technical issues should be anticipated so that the report and
consultant answer questions regarding plume extent and regulation
before the report is issued.

It has been the authors' experience that the maps and cross sections are typically the most reviewed portion of the data. Usually
the presentation format is similar to the usual geologic and civil
engineering presentations adapted for use in contaminant hydrogeology. The maps then should be carefully prepared, and the data
must be accurate since maps and sections are sometimes reviewed
without a careful text perusal. Although the text is the most complete description of what was done, the maps and sections are
shortcuts used by all interested parties—consultants, regulators,
and clients. There is always the possibility of these documents
being used out of context.

Each investigation is different—the type of problem and geology
will always differ from site to site. Even though similarities may exist,
one can never apply the same solution to different sites and assume
one can obtain equally effective remediation. The information derived in the investigation and testing is valid for that unique site and
must support site-specific conclusions.

Comments on Using Previous Investigation Reports

When conducting or working an investigation that another consultant has done, one should be careful to be sure that the information is accurate and agrees with your work. Often you must refer to reports done by other geologists or consultants and agencies and must evaluate the validity of the work as it relates to your site. If you accept work that is inferior or flawed, you will compromise the accuracy of your work. Several problems may arise in using other consultants' work. The following examples are by no means a complete list, and the hydrogeologist must always guard against accepting dubious information.

Problems can arise from an incomplete definition of the extent of vadose contamination. If this is the case, a residual contaminant may remain that will leach additional contaminants into the groundwater body, causing future problems if the contamination remains in the ground for some time, or additional costs when remediation is attempted. The regulating agency may ask for more definition work and remediation. Obvious cost and legal problems will arise due to incompleteness of a report.

Another series of problems arise from cross-contaminated aquifers as the result of previous well installations. When wells (or exploratory borings) connect several water-bearing horizons, the potential for interconnection exists. The strata may not be saturated when drilling, yet when seasonal precipitation recharges the dry strata, possible cross connection could occur. Contaminated water may move vertically in the interconnection, spreading the problem. This may occur if the lithologic logging was incomplete, sloppy, or simply not done, as is the case in many agricultural wells. Logging styles vary given the original purpose of the boring, and useful hydrogeologic information may not be included in an inferior well log. Contact identification may be vague or only estimated while drilling, so an older well log may be inaccurate. The well design detail should be reviewed to check to see if the designed screened interval is contiguous with the supposed aquifer and sealed into the aquiclude. If the construction of the well is sloppy, seals and sand packs may not be located at the intended design interval.

Another problem is incomplete definition of the "zero line," or the furthest extent of the groundwater plume. If the extent of the plume is not defined by chemical data, regulating agencies will require more work and add cost to a project. More importantly, an incomplete definition of the plume may skew the remediation effort, and not completely capture the contaminants. Thus, portions of the plume may continue to migrate, creating complex cleanup and liability problems.

A final problem involves the quality of all chemical analytical data, especially that from previous investigations. The data may be suspect if sampling and quality assurance and control procedures were not properly accomplished. Sampling and sample-handling procedures should be reviewed in light of the reported analytical values. The recorded analytical data should be reviewed in light of the laboratory procedures for accuracy and reproducibility. Inferior chemical data may lead to erroneous interpretations and conclusions, which may compromise existing efforts and the direction of the next phase of the study. Good chemical data are crucial for placement of exploratory borings and monitoring wells used to delineate the plume.

EXAMPLE APPROACH TO A SITE INVESTIGATION

You are retained to investigate an underground tank site. A previous consultant performed a preliminary site assessment, which revealed petroleum hydrocarbons in site soil samples. A "grab" groundwater sample revealed both hydrocarbons and solvents present in the water. The site owner swears he never used solvents onsite. Since he has already had some investigative work done, his money is short, but he was sent a letter from the regulatory agency requiring him to investigate the extent of the problem. How do you proceed?

The three previous borings were placed near the suspected source, the underground tanks (see Figure 15). Since the petroleum hydrocarbons are in soil and groundwater, the investigation must find the vertical and horizontal extent of the hydrocarbon contaminants. Solvents apparently are also present based upon grab water samples collected from the uncased borings. Monitoring wells must be placed so that the extent and source (if present) of the solvents are known. The exploratory borings and well placement in Figure 15 were chosen to allow as broad a site coverage as the budget allowed.

Note that monitoring wells cost more than borings but leave permanent monitoring points. Borings, however, allow soil sampling without necessarily building wells. Exploratory borings should cover the site in known contaminated areas for vertical and horizontal vadose zone coverage. Wells are both near and distant from the source to

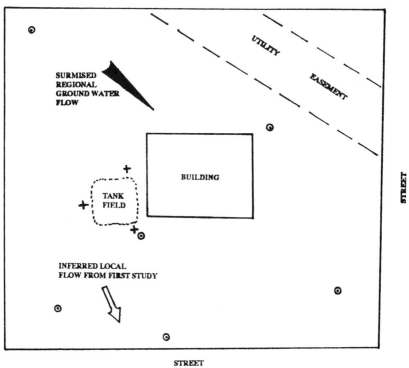

GARAGES AND INDUSTRIAL BUILDINGS

SURMISED
REGIONAL
GROUND WATER
FLOW

UTILITY

EASEMENT

STREET

BUILDING

TANK
FIELD

INFERRED LOCAL
FLOW FROM FIRST STUDY

STREET

+ PREVIOUS BORING LOCATION

⊙ PROPOSED WELL LOCATION

Figure 15.   Example of well locations for defining and monitoring a plume.

ascertain plume extent and search for potential solvent sources. If solvents were never used onsite, the best guess is then an upgradient source, so one well was placed there.

The exploratory boring program will sample soil in all boreholes to search for the hydrocarbon contaminants. Selected soil samples will be analyzed for solvents in lesser quantity since some physical evidence precluding solvent presence in soil will be required by the regulators. However, analyzing all samples is not warranted, given the site history and the available budget (so a data/budget compromise is struck).

Soil samples from the aquitard underlying the aquifer will be retained for analysis of solvents as well as the permeability test so that the depth of contaminant penetration can be established. Chemical analysis of the hydrocarbons will not be done since, given their chemistry, these contaminants should not have affected the aquifer (another data/cost compromise).

The six monitoring wells will be sampled for both hydrocarbons and solvents. Since the previous data showed the presence of both, this should confirm whether solvents are present or whether there was a lab or sampling error at the time of the previous investigation. The monitoring well sampling protocol should be reviewed for completeness so that questions of improper sampling do not compromise data.

This approach should give the hydrogeologist the basic information required to form valid conclusions. The geology, aquifer, and aquitard should be identified as well as groundwater occurrence and flow direction. The chemistry of soil samples should yield the minimum extent of vadose vertical and horizontal contamination. Groundwater data should show the contaminants distribution onsite, with some indication as to whether the contaminants have moved offsite. The upgradient well should answer whether solvents are moving onsite from an offsite source. The report should accurately and concisely report all information, using the appropriate technical standards and formats required by the regulating agency.

Unanswered questions will remain following the investigation, since only rarely are all answers found from a single study. Another phase of study may be required so that the hydrogeologist feels the coverage is adequate. The hydrogeologist must give the client the best possible opinions and most scientifically accurate report and interpretation of the data, subject to budget availability and legal requirements. This also includes delivering the "bad news" if necessary, which may mean spending more money. The client may not wish to spend additional money. Legal implications of conducting more investigation are ultimately the client's responsibility. Although the hydrogeologist may have done as much as possible, the data may not be complete and the client may elect not to budget additional work.

CHAPTER 8

Aquifer Analysis

INTRODUCTION

Aquifer analysis for contaminant hydrogeology investigations is aimed at hydraulic control of the plume. Groundwater is extracted for surface treatment and discharge. Aquifer properties can be calculated by measuring water level drawdowns and recoveries in a pumping well and selected observation wells and matching graphed data plots to type curves based on theoretical predictions or use in numerical formulas. Background information should be obtained and reviewed to achieve the best possible data before a test is scheduled. Data review should include (as a minimum)

1. well construction details
2. location of the proposed pumping well(s) and observation wells
3. review of subsurface lithology to evaluate suspected or known aquifer extent and thickness, aquifer/aquitard relationships
4. the anticipated degree of homogeneity and isotropy of the aquifer to be tested

AQUIFER PUMPING TEST

Pumping tests are performed to estimate specific aquifer properties such as transmissivity, storativity, hydraulic conductivity, and potential capture zones, and to estimate well efficiency. With few exceptions, most aquifers will not be laterally extensive or isotropic; therefore, subsurface conditions must be anticipated before a test is set up. Nonuniform geologic conditions commonly encountered in the subsurface that will influence test results are discussed below.

Potential Boundary Conditions

Are there recharge or discharge areas close by? Are there known or suspected impermeable or less permeable geologic features within the test area? These conditions may significantly impact a pump test (i.e., will influence the rate of cone development and shape of the cone of depression after pumping begins). If water level data through the course of a test changes suddenly, a boundary condition may be the cause.

Delayed Drainage

Typically, many areas of investigation will not be homogenous. Stratigraphic variability, particularly where an aquifer is composed entirely or in part of fine-grained materials (e.g., silts and clays), causes a natural resistance to the vertical movement of water under the influence of gravity due to such properties as cohesion and adhesion. The results of delayed drainage are recognized on field data plots where the drawdown curve flattens and then steepens again. This phenomenon can best be detected during the longer-duration constant-rate discharge test.

Secondary Permeability

Joints, fractures, vesicles, and solution cavities may not be uniform and/or continuous (e.g., metamorphic or igneous rocks, or karst topography). An observation well very close to the pumping well may experience little or no drawdown during a pump test simply because there is little or no hydraulic interconnection in the subsurface materials between the two wells. Typically, jointed, fractured, and karst terrains are the most difficult geologic areas to obtain good pump test data and make interpretations.

Aquitard Leakages

Recharge to a pumping aquifer through overlying and/or underlying confining layers (aquitards) can be suspected when water levels in wells in the pumping aquifer increase after the pump test has been running (usually several hours into the test). Leakage can be assumed if no obvious recharge sources are identified.

Underground Utilities

If a pump test is being performed in shallow, near-surface wells in the area of an old water or sewer line, or where storm and sanitary sewer systems exist, inflow or outflow conditions from old pipes may potentially influence a pump test and result in data uncertainty. It is a good idea to check and see if underground utilities are present in the test area and try to find out their depth, and if possible, present condition (i.e., contact local city or county agencies).

Insufficient or Poor Well Development

Absolutely essential to the success of any pump test is the integrity of the pumping well and observation wells. The pumping well and observation well construction and development is extremely important. If the wells are not properly developed before a test, typically early data (usually several hours) may not be representative or may appear unexplainable (erratic). In many cases, what this early test data represents is well development during the early stages of a test. It is highly recommended that well development precede a test to enhance data collection, particularly early test data.

Pump Test Equipment

The amount of equipment necessary to conduct a pump test can be quite extensive and expensive, depending upon the type or types of tests to be performed, site logistics, well design, presence and type of contamination, and number and location of observation wells. The following equipment list is typical but can vary depending upon test requirements:

submersible pump (stainless steel components preferred; equipped with a check valve and appropriate control box)

pump discharge line or piping (compatible with discharge pump)

pipe dog or tripod (to secure pump in well)

flow valve control (gate or ball valve) and flow meter system

discharge piping or line (flat-lying flex hose is good)

power source (AC power or suitable portable generator)

stop watches (a minimum of 2)

water level measuring devices (electric sounder, steel tape, oil/water interface probes)

data logger and pressure transducer

semilog and log-log graph paper

rulers, french curves, mechanical pencils

scientific calculator

site safety equipment (tyvek suits, gloves, boots, etc.) based on known or suspected site conditions. Equipment should be outlined in a site-specific safety plan.

data sheets to record drawdown (recovery data, time, well numbers, pumping rates, etc.)

tools, fittings, etc., as required

lighting (flashlights and/or lanterns)

In addition to field equipment, the hydrogeologist, geologist, engineer, or technician should bring the following documentation into the field during a test (copies only—the originals stay in the investigation files):

boring logs for wells to be pumped and monitored

well completion details of wells to be pumped and monitored

well development data sheets (if available)

groundwater sampling purge data

latest chemical data for groundwater samples collected from the monitoring network

Pretest Field Procedures

1. Obtain a topographic and/or facility map if available. Review lithologic logs and well construction details.
2. Site visit: Identify access constraints and potential traffic problems, and inspect integrity of wells to be tested. Identify potential or required discharge points and distances from test wells.
3. Secure all required permits, right-of-ways, variances, and so forth from proper state or local agencies, or private parties.
4. Obtain and review pump histories for wells to be tested. For example, if the wells were developed, try to estimate discharge rate. Also, review sampling records, if available. Purging data can be very useful to estimate well yield(s).
5. Assemble all necessary equipment to perform the pump test. Make

sure that all equipment is functional before you go to the field and begin a test. Decontamination of equipment is essential on all hazardous waste sites.

Pretest Well Monitoring

Selected pumping wells and "background" or observation wells should be monitored for at least one day (24 hr) prior to a test. The objective of pretest well monitoring is to identify changes in water levels due to diurnal changes such as tides, effects of irrigation, or pumping from local domestic or municipal wells. Graphic plots of water level changes versus time are very useful in identifying water level fluctuations unrelated to the pump test. Water levels can be periodically measured by hand or by using a transducer/datalogger system to record data prior to a test.

Field Procedures

1. Measure static water level in the discharge well. Measure depth to bottom of well casing, and record depth to accumulated fines or possible well obstructions. Calculate available water column.
2. Install submersible pump in the well to the desired depth of placement. Usually the pump intake is placed opposite the bottom of the screened interval (placement will vary according to well construction and geologic conditions). It is a good idea not to set the pump on accumulated silts or clays in the bottom of the well because of the potential for pump damage. Ideally, test wells should be properly developed before a pump test.
3. Observation wells should be identified that may be influenced during the test, and water levels in the observation wells should be measured and recorded. Periodically through the test, water levels can be measured in the observation wells to evaluate radius of influence and development of the cone of depression produced by pumping. Remember to compare data with background data to make sure changes in observation wells are related to pumping and not caused by natural phenomena.
4. Field measurements to be taken during the test include the following:
   a. time since pumping started
   b. time since pumping stopped (recovery)
   c. depth to water
   d. discharge rate
5. The greatest fluctuation in water levels typically occurs at the beginning, and sometimes at the end, of a test. Therefore, water level

Table 1. Suggested Measurement Intervals

Elapsed Time	Water Level Measurements
0–5 min	Every 0.5 min
5–15 min	Every 1.0 min
15–40 min	Every 5.0 min
40–end of test	Every 10.0 min

Note: The above time intervals are guidelines. Readings may vary according to site conditions.

measurements at the beginning of a test should be recorded more frequently. Ideally, a datalogger/transducer system should be used to collect drawdown data. At a minimum, depth to water and the discharge rate in the pumping well should be measured at the time intervals shown in Table 1.

6. Additional remarks:
   a. Discharge rates for a test must be maintained within 10%. Check discharge rate frequently.
   b. Identify reference point from which water levels are measured (e.g., top of well casing, top of well box).
   c. Plot data during every test. Data plots are very useful in monitoring a test and assist in determining when a test should be terminated (e.g., rapid dewatering of the well casing). This is very useful if the electronic memory of a datalogger fails.
   d. Do not change water level measuring devices during a test, unless required due to equipment failure. Always have a backup measuring device. If you assume you won't need one, you probably will (Murphy's law!).
   e. Be sure to note the time the pump was activated, and when the pump was shut off. Also, be sure a check valve has been installed to prevent water from running back down into the well. Without a check valve, essentially you are recharging a well artificially after the pump is turned off.
   f. If the test well or wells are part of a monitoring network and have historically contained chemicals, decontaminate equipment before every test, between tests, and after the last test.
   g. Cleaning equipment after a test is a good policy to adopt to ensure that someone else does not use contaminated equipment on their jobs and that you do not use contaminated equipment on your job.

Table 2.   Well Recovery Measurement Intervals

Elapsed Time	Water Level Measurements
0–5 min	Every 0.5 min
5–15 min	Every 1.0 min
15–40 min	Every 5.0 min
40–end of test	Every 10.0 min

Note: Water level measurements can be terminated if less than 0.01 ft of change in water level is measured over a 30–60 min period. Unchanged readings for this period of time typically indicate that recovery to original static water level will take a long time. A pump test can be terminated if recovery occurs to within 80–90% of static water level.

Well Recovery Test Field Procedures

1. Measurements of water levels begin immediately after the pump has been shut off. Measurements should be made at the time intervals shown in Table 2 as a minimum.
2. Do not be alarmed if a water level returns to a depth shallower than (above) original static level as measured from ground surface. This may be the result of normal diurnal changes or a "rebounding" effect due to pumping. Often, a water level will rise slightly above the original static level, then fall again close to the original static level.
3. Plot recovery measurements (drawdown versus time) during well recovery.
4. Under no circumstances should the pump or any down-well test equipment (e.g., transducers) be removed from a well until the recovery test has been completed. Removal of equipment will give erroneous water level data.

SLUG TESTING

Both transmissivity and hydraulic conductivity of an aquifer can be estimated from the rate of rise or fall of the water level in a well after a slug of known volume is either instantaneously introduced or removed from the existing water column. The main advantages of performing slug tests as opposed to pumping tests are the following:

1. They are less expensive to perform.
2. Less equipment is needed.
3. It takes less time to obtain data in the field.
4. Data interpretation/reporting time is shorter.

5. They can be used where pumping data may not work (e.g., low-yield conditions).
6. They can be used in small-diameter wells.

On the other hand, the disadvantages include the following:

1. T and K values are estimations, at best, in most cases.
2. The aquifer is not stressed sufficiently to properly evaluate the area of influence from tested wells, usually just beyond the sand pack.
3. They are only applicable to low-yield aquifers.
4. They are not applicable to large-diameter wells.
5. They can yield very erroneous data if the test well is not properly developed.

Equations developed by Ferris and Knowles (1954); Cooper, Bredehoeft, and Papadopoulos (1967); and Bouwer and Rice (1976) are commonly used methods to evaluate slug test data.

Ferris and Knowles (1954)

$$T = \frac{114.6 \; V \; 1/t;}{s} \quad K = \frac{T/b}{7.48}$$

where T = transmissivity (gpd/ft)
    V = slug volume (gal)
    $1/t$ = selected reciprocal of time data point
    s = drawdown or residual drawdown (ft)
    K = hydraulic conductivity (ft/day)
    b = aquifer saturated thickness (ft)

Cooper, Bredehoeft, and Papadopoulos (1967)—Modified

$$T = \frac{X \; (r_c^2) \; (7.48);}{t} \quad K = \frac{T/b}{7.48}$$

where T = transmissivity
    X = selected Tt/r_c^2 value
    r_c^2 = radius of well casing squared (ft)
    t = x-axis time value corresponding to selected Tt/r_c^2 value (in days)
    K = hydraulic conductivity (ft/day)
    b = aquifer saturated thickness (ft)

Bouwer and Rice (1976)

$$K = \frac{r^2 \ln (R_e/R_w)}{2L(t)} \frac{\ln \ H_o}{H_t}$$

where   K = hydraulic conductivity
R_e = effective radius (ft)
R_w = borehole radius (ft)
L = screen length (ft)
t = time of residual drawdown measurement
H_o = instantaneous drawdown at time t = 0
H_t = residual drawdown at time = t
ln = natural log

$$T = Kb$$

where   T = transmissivity (gpd/ft)
K = hydraulic conductivity (ft/day)
b = aquifer saturated thickness (ft)

STEP DRAWDOWN/WELL RECOVERY TEST PROCEDURES

The purposes of a step drawdown/well recovery test are to

1. estimate aquifer transmissivity
2. select the optimum long-term discharge rate for a constant-rate discharge test
3. identify wells in hydraulic communication with the pumping well for monitoring during the constant-rate discharge test

A step test can be completed within a relatively short time span (usually 6 to 10 hr), and only a single well is needed for the estimation of transmissivity. In comparison, a constant-rate discharge test typically requires a minimum of 24 hr of pumping from the pumping well and at least one observation well within the anticipated radius of influence to calculate transmissivity and storativity. If the saturated aquifer thickness (t_o) is known, hydraulic conductivity can be estimated. The step drawdown/well recovery test consists of two phases: the step drawdown test, and the recovery or residual test.

Step Drawdown Test

The step drawdown test involves pumping a well at an initial discharge rate (Q_1), which is incrementally increased (stepped) while drawdown (s) is measured in the test well at various time intervals. Ideally, the step drawdown test utilizes three or more different discharge rates or steps (Q_1, Q_2, . . .) at a minimum, with each subsequent flow rate increased from the previous flow rate. Under most conditions, rate increases will vary. Therefore, particular attention to subsurface geologic conditions, well design, and available water column in the well is critical to estimate potential flow rates before a test is started. Typically, steps are doubled, if possible. However, increases must be compatible with site subsurface conditions. An initial "conservative" flow rate is recommended to evaluate aquifer response and material capability.

The duration of a particular step depends on several factors. The observed water level with respect to available water column in the pumping well dictates the length of a step and potentially how many steps a test will have. The target duration of each step should be at least 60 min, so that enough data points can be plotted to establish a trend on the semilog graph. Sixty minutes is preferable unless drawdown conditions (e.g., dewatering) preclude a step from running that long. If the water level within a step does not level off or begin to show signs of equilibrating, you may be looking at the last step in the test. If this occurs in the first step of a test, you may have to shut the pump off, allow the well to fully recover, and start the test again at a lower pumping rate. Or it may mean that a step test cannot be properly performed due to low-yield conditions. Likewise, if very little drawdown occurs in a step and no changes in drawdown occur over a 35-min period, proceeding to the next step is usually justifiable.

The selection of the discharge rates for each corrective step must consider the water level in the discharge well observed during the previous step. It is better to select a rate increase that is conservative than to overstress the aquifer and end up dewatering the well. The effects of pump rate increases for different steps can be tracked by constructing a simple graph before a step drawdown test is conducted. The graph should include (1) the available water column present in the well, (2) well design, (3) aquifer/aquitard or aquiclude relationships, and (4) location and depth of the submersible pump intake in the test well.

Generalized Discharge Rate Criteria

1. If drawdown levels off at less than 25% of available water column, flow rate can be increased.
2. If drawdown levels off between 25 and 50% of available water column, flow rate may be increased. Use your judgment on expected drawdown based on observed drawdown from previous step.
3. If drawdown levels off between 50 and 75% of available water column, flow rate may or may not be increased. Again, look at previous data and perhaps select a conservative increase in order to prevent dewatering.
4. If drawdown is greater than 75% of available water column, a flow rate increase will most likely dewater well. A conservative pump rate increase may be possible but should only be attempted if previous data indicate such an increase might be possible.
5. If drawdown is greater than 90% of water available water column and dropping, prepare for recovery test. Shut pump off and begin recovery test.

The above criteria are meant to be only guidelines. Obviously, each test site will vary according to hydrogeological conditions — aquifer/aquitard relationships, well design, and the amount of water available in an aquifer. A good rule-of-thumb is to increase steps conservatively if you are not sure how the pumping well will respond to an increase in the discharge rate. Extra steps are preferable over not enough steps to evaluate potential well yield and estimate transmissivity.

In some instances, if no increase in the flow rate during the step drawdown test can be sustained by the pumping well, it is recommended that the current flow rate be maintained for the duration of the test, as long as you have data for at least three steps. If dewatering is inevitable and you want several steps for a test (for example, a constant-rate test is not going to be performed), shut off the pump, allow the pumping well to fully recover, and start the test over at a lower discharge rate.

Residual or Recovery Test

The residual or recovery test involves monitoring well recovery after the pump has been shut off. Recovering water level measurements are made at various time intervals until original static water level has been reached. If a test is being performed in an area where low-yield aquifers are being tested, well recovery may take 24 hr to fully recover, or longer. Typically, in a low-yield aquifer, 80–90% recovery may be acceptable. The last 10–20% of recovery usually takes the longest to achieve, and it may not be cost effective to run a test out that long. Ideally, recovery data should be collected for 2–4 hr (depending upon specific site conditions).

Remember: only qualitative evaluations of well production can be speculated from recovery, since the discharge rate was not held constant during pumping.

Evaluating Aquifer Test Data

Theis (1935)

Theis noted that during a pump test in a well that penetrated an extensive confined aquifer, if the discharge rate is held constant, the area of influence increases with time. He was able to conceptualize the formation of the cone of depression in this type of subsurface environment and realized that the rate of water level decline multiplied by the storativity and summed over the area of influence equalled the discharge rate (Kruseman and DeRidder, 1976). Theis also believed that as long as water was continually being removed from storage in an isotropic aquifer, drawdown will continue over time and that, theoretically, no steady-state flow would exist. Theis' nonsteady-state equation is as follows:

$$W(u) = 0.5772 - \frac{\ln u + u}{2.2!} - \frac{u^2}{3.3!} + \frac{u^3}{4.4!} - \frac{u^4}{+} \cdots$$

The Theis equation is an experimental integral for the "well function of u."

The values of W(u) vary as u varies. The Theis equation can be rewritten in a simpler form:

$$T = \frac{114.6\ Q\ W(u)}{s} \quad \text{and } S = \frac{Tut}{1.87r^2}$$

where T = transmissivity of aquifer (gpd/ft)
        S = storativity of aquifer (dimensionless)
        Q = pumping discharge rate (gpm)
        s = drawdown in cone of depression (ft)
        t = time since pumping started (days)
        r = distance to pumping well center (ft)
  W(u) = "well function of u"; an exponential integral (infinite series)

The Theis equation follows Dupuit's assumptions:

1. The aquifer is seemingly infinite in areal extent.
2. The aquifer is homogeneous, isotropic, uniformly the same thickness, and horizontal.
3. Prior to pumping, the potentiometric surface is horizontal (i.e., flat).
4. The aquifer is pumped at a constant discharge rate. The pumping well is 100% efficient.
5. The pumping well fully penetrates the aquifer, thereby receiving water from the entire thickness of the aquifer.
6. There is no induced recharge during the pump test.
7. Water is discharged instantaneously with drawdown.
8. The potentiometric surface has no slope.

Type Curve or Match Point Method—Theis Method (Figure 1)

1. Take W(u) versus 1/u or u curve and superimpose over s versus t field data curve.
2. Make sure both horizontal and vertical axes of data plot and type curve are parallel.
3. Select a "match point" (preferably on an even log scale:10, 100, or 1000). Avoid early data matching. Later data are preferable.
4. Match point gives you four coordinates:
   a. W(u)
   b. 1/u or u
   c. s
   d. t
5. Calculate the lower integral equation:

$$u = 1.87 \ r^2 \ S/Tt$$

To get u, take 1/u and use (u):

$$T = 114.6 \ Q \ W(u)/s \qquad (1)$$

and

$$S = Tut/1.87r^2 \qquad (2)$$

To solve Equation 1:
a. Get Q from test.
b. Get s from match point.
c. Get u from match point (reciprocal of 1/u).

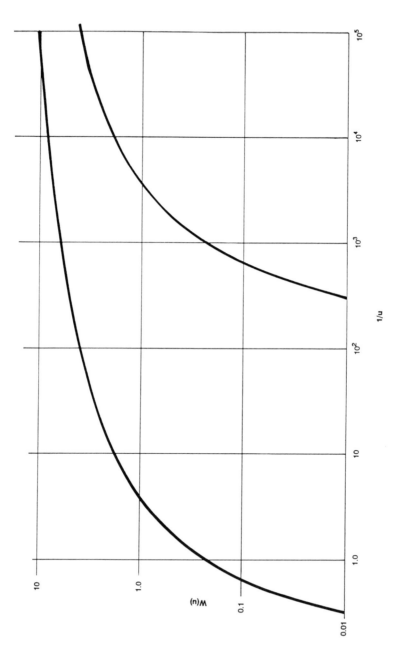

Figure 1. Theis method: nonequilibrium type curves.

To solve Equation 2:
a. Get T from Equation 1.
b. Get u from match point.
c. Get t from match point.
d. Get r from field measurements (distance between pumping well and observation well in feet).
Note:
a. t is always expressed in days.
b. Data is plotted on the same scale graph paper as the type curve.

Example. Assuming the following parameters:

$$T = 62,500 \text{ gpd/ft}$$

$$S = 2 \times 10^{-5}$$

$$Q = 250 \text{ gpm}$$

$$r = 750 \text{ ft (distance from pumping well)}$$

$$t = 1 \text{ day}$$

calculate drawdown (s), 1000 ft from the pumping well for one day (modified after Driscoll, 1986):

$$u = 1.87 \ r^2 S/Tt \ (t \text{ is in days})$$

$$s = 114.6 \ Q \ W(u)/T$$

Solve for u:

$$u = 1.87 \ r^2 S/Tt = \frac{1.87 \ (750)^2 \ (2 \times 10^{-5})}{62,500} = 3.4 \times 10^{-4}$$

and find W(u) using tables:

$$W(u) = 7.4097$$

Now you can solve for s, using the rearranged Theis equation:

$$s = \frac{114.6 \ Q \ W(u)}{T} = \frac{114.6 \ (250) \ (7.4097)}{62,500} = 3.4 \text{ ft}$$

In addition to plotting pumping data, recovery data (pumping well and observation wells) or what is known as "residual drawdown" can be used to calculate transmissivity and storativity. Residual draw-down versus time since pumping stopped is plotted on log-log graph paper. The Theis curve is used to find the match point, and T and S values are found using the two Theis equations. Remember: storativity can only be calculated if data from an observation well is used.

Cooper and Jacob (1946)

The Cooper-Jacob graphical method for evaluating transmissivity (T) is based on the Theis equation, although the conditions for its application are more restrictive. For values of u ≤ 0.05 (usually u is greater than 0.05 early in a pump test), the modified nonequilibrium equation will give the same results as the Theis equation (Kruseman and DeRidder, 1976). The straight-line method is essentially the same as the Theis equation, except that the exponential integral function W(u) is replaced by a logarithmic term. The equations for transmissivity and storativity are as follows:

$$T = \frac{264Q}{\Delta s} \qquad (3)$$

$$S = 2.25 \, Tt_0/r^2$$

where
T = transmissivity of aquifer
S = storativity of aquifer
Q = pumping discharge rate
t_0 = time at zero drawdown intercept
r = radius of the pumped well
Δs = change in drawdown slope across 1 log cycle

Cooper-Jacob Straight-Line Method

For nonleaky, confined aquifer analysis, the Jacob's method can be used to calculate T and S, following the steps listed below:

1. Plot pump test data time (t) versus drawdown (s) on semilogarithmic graph paper (drawdown on arithmetic scale and time on log scale) from observation well data.
2. Draw best-fit straight line through data points and extend line up to zero drawdown (S = 0)
3. Measure slope of the straight line (Ds for one log cycle — slope is defined simply as "rise over run").
4. Read value of t_0 at intersection of zero drawdown line. Value read

will be in minutes. This value will have to be converted to days to estimate storativity.

5. Solve for Equation 3 to get transmissivity (T).
6. Use observation well data (drawdown and distance to pumped well) to calculate storativity (S). Use T value from Equation 3.

The Cooper-Jacob data plot should be used on every pump test to track the progress of the pumping well and impacted observation wells. Typically, drawdown and time data points are taken from a datalogger (pumping well and close proximity observation wells) and plotted on semilog graph paper. Later in the test, a best-fit straight line can be used to estimate transmissivity and used with observation well data to estimate storativity of the aquifer.

Although many regulatory agencies are interested in the "blackbox" data analyses of computer models, supplementing these data with hand-calculated data is important for three specific reasons:

1. Hand-calculated data should be used to support or refute computer model calculations.
2. By performing hand calculations, you are demonstrating to regulators and potential peer reviewers that you understand the mechanics involved in quantifying aquifer properties.
3. Pump test data are used to select, develop, and implement remediation measures. Remedial action is a BIG dollar item in completing a project; therefore, you may not feel comfortable with using "blackbox" computer analysis data alone. Remember: every computer model can generate T, S, and hydraulic conductivity values; however, you alone must determine whether they are representative of site-specific conditions. It is advisable that both hand-calculated and computer-generated aquifer parameters be compared with collected field data to verify that they make sense for the "system" you are working in.

Distance-Drawdown Method

When several observation wells are monitored during a constant-rate discharge pump test, the distance-drawdown method of aquifer analysis can be used to calculate transmissivity and storativity. Three observation wells are required as a minimum to use this method of analysis. Drawdown data from three observation wells recorded at the same time can be plotted on semi-log graph paper—drawdown in feet on the arithmetic axis and distance from the pumping well in feet on the log axis. The relationship of drawdown to distance from the pumping well results in a straight line when plotted on semilog paper.

The points on the semilog plot are connected, and the line is extended to the zero drawdown line, as with the best-fit line used for the Cooper-Jacob method. The intercept at the zero drawdown line is the r_o value.

The modified nonequilibrium equation can be transformed to use distance-drawdown plots to calculate (T) and (S):

$$T = \frac{528Q}{\Delta s}$$

and
$$S = 0.3\ Tt/r_o^2$$

where   T = transmissivity of aquifer (gpd/ft)
          S = storativity of aquifer (dimensionless)
          Q = pumping discharge rate (gpm)
          t = time since pumping began (days)
          r_o = zero drawdown intercept of extended straight line
                 through a minimum of three data points (ft)
          Δs = change in drawdown slope across 1 log cycle

As is the case with previously described pump test methods, Dupuit's assumptions are believed to be true for this analysis method.

In addition to being able to calculate T and S using the distance-drawdown method, well efficiency can also be estimated (Driscoll, 1986). The well efficiency values (percents) obtained from this method are estimations at best. Several factors come into play that directly affect this calculation:

1. *well design:* type and slot-opening size, length of well screen (i.e., fully or partially penetrating screen), compatibility of artificial filter pack with screen slot and native aquifer material
2. *subsurface geology:* aquifer material (heterogeneity) and its effects on flow to the well, potential recharge or discharge sources (e.g., boundary conditions), degree of saturation of aquifer (fully or partially saturated), aquifer too fine-grained to adequately determine well efficiency, leakage from aquitards
3. *pump test parameters:* pump rate too large to effectively evaluate aquifer (i.e., turbulent flow interferences as a result of artificial filter pack bridging), pump rate too small to effectively stress aquifer, discharge lines too restrictive to measure actual discharge rate, fiction losses through discharge pipes and hoses reduce observed discharge rate
4. *well development:* improper well development prior to the pump test (e.g., drilling fluids still in aquifer material and artificial filter pack; artificial filter pack shifting during pumping as grading occurs)

CHAPTER 9

Remediation and Cleanup

INTRODUCTION

The goal of site remediation is to restore soil and groundwater quality to precontamination conditions — ideally, to restore the site to "natural" conditions in which all contamination is removed. In most cases, groundwater cleanup objectives are directed toward an aquifer nondegradation goal, where the groundwater is suitable for all beneficial uses. The intent of government regulations are to restore the site to preexisting conditions. Attaining this goal is almost never realized since it is impossible to be absolutely certain that every molecule of contamination is removed.

The lead agency (local, state, or federal) overseeing the site will usually establish cleanup standards or groundwater protection standards, following a review of all the data and negotiations regarding the site in question. The standards are based on numerous criteria, including the site information, contamination type and extent, potential threat to human health, and protection of future soil and groundwater quality. Agencies will usually lean to conservative standards, although they are almost never set to the most strict standards (as, for example, the recommended maximum contaminant levels). The cleanup standard often considers the allowable contaminant concentration left onsite, which would not adversely affect human health, or soil or groundwater quality.

CONCEPTUAL APPROACH TO SITE CLEANUP

Site remediation considers the limits of technical expertise, existing government regulations, environmental protection desired, and economics. Costs for site remediation can quickly escalate into considerable sums. The remediation plan and execution of the cleanup are generally the most costly and time-consuming portions of the project, given the complexities of subsurface conditions. As the cost of site cleanup has risen in the past few years, questions are being raised concerning what the governing philosophies should be. Considerable progress has been made recently in regulation, engineering practice, environmental awareness, source control, and waste management procedures. This progress enables more reasoned decisions to be made regarding contaminant migration and threat to human health. The EPA and some states have prioritized sites for cleanup so that contaminated sites posing potentially extreme health hazards are addressed first.

The most frequently asked cleanup question is, How clean is clean? In other words, what is the appropriate residual contaminant concentration to be left in soil or groundwater, and how much effort is needed to clean the site while safeguarding environmental concerns? As chemical analytical techniques become more precise, smaller concentrations of contaminants are measured. Remediation goals for cleanup may address "natural conditions," but is this a realistic goal? Residual contaminants trapped in the vadose zone or aquifer porosity may be extremely difficult or technically infeasible to remove. Additional toxicological studies of compounds may differ in what concentrations of compounds are "toxic." If laboratory standards are enforced as cleanup standards, are they realistic to protect soil and groundwater? The costs of these decisions are factored into the site remediation since the responsible party will have to perform cleanup (see Figure 1). The cleanup standard will determine the level of effort and budget required for remediation.

Clearly, site remediations may—and have—bankrupted responsible parties, and subsequently the government must complete the task. The financial resources of governments or industry are not limitless. Cleanup costs must be paid by someone, and the costs and number of cleanup sites continue to increase. Management and allotment of funds must be considered carefully, or the cleanup will stall and languish, defeating the purpose of the remediation. Negotiation delays may allow the plume to move and the problem to expand.

Site remediation is a compromise, using reasonable judgment, based upon the available information, applicable regulations, and cost effectiveness of the cleanup. Sites will rarely, if ever, be cleaned to preexisting or natural background conditions. Often, inflexible positions taken by government, potentially responsible parties, and environmental lobbying groups delay cleanup efforts.

REMEDIATION AND CLEANUP

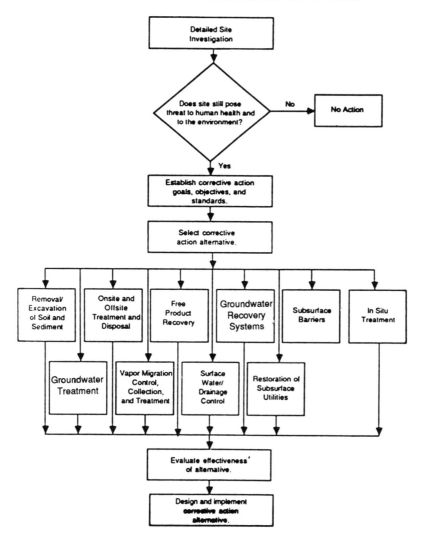

Figure 1. Flow chart of corrective action for an underground tank leak. *Source*: EPA (1985).

Although the intentions may be good, the time delay allows the contaminants to migrate, expanding the potential cleanup area, while the interested parties cling to unreasonable or utopian positions.

CLEANUP TECHNOLOGIES

Numerous technologies to remediate sites exist. Some technologies have more applications than others and are more specialized or contaminant specific. The EPA has reviewed these technologies and often refers to them as the "best available technology" (BAT) for cleanup cost. The determination of the effectiveness of a cleanup technology may be somewhat subjective, and sometimes the technology may not be acceptable in certain areas — for example, use of air stripping is rapidly being discouraged in southern California since the transference of the contaminant from water to an existing air pollution problem is no longer acceptable to the local air management agencies.

It is important to remember that there is no magic formula for remedial solutions. The following discussion is not meant to be an exhaustive review of all available cleanup methods and design approaches. No technology is suitable for all sites since every site and contamination situation is different. New technologies are being developed and tested; also, several technologies may be used together, such as a groundwater capture and extraction system combined with soil vapor venting. Some technologies are specific to contaminant or may have site geology, cost, or regulatory limitations. A long time period may elapse before obtaining remediation plan approval, so interim remediation may be required to prevent further expansion of the contaminant plume. The remedial approach must be designed for the site and contaminant in question to yield the best results.

Cleanup technologies have been reviewed by EPA information and source documents and have been described in the literature (see, for example, EPA, 1987b). All technologies must consider required use, construction, and discharge permits. A brief review of some of the more common BATs in general use listed by EPA and used for groundwater are considered below.

Excavation

Excavation involves the physical removal of the contaminated materials for disposal at a hazardous waste or other disposal landfill site. Excavation is being discouraged by newer federal regulations that favor alternative waste treatment technologies at the contaminated site. Municipal refuse will continue to be placed in landfills for some time to come. Consequently, sanitary landfills will be needed for the indefinite future for domestic refuse, which may contain small quantities of hazardous materials. Attempts are being made to segregate and remove hazardous materials from domestic refuse to keep the sanitary landfills from becoming de facto hazardous waste landfills. Considerations for any excavation project include the volume of material to be removed, equipment to be used, source and type of clean backfill, compaction specifications, and shoring for sidewall stability. Excavation of hazardous waste for disposal is costly, and the cost of disposal and transportation to the landfill must be included. Finally, postexcavation sampling and testing are required to determine the effectiveness of the excavation cleanup.

Air Stripping

Air stripping uses the volatilization characteristics of the contaminant to separate it from groundwater. The contaminated water is pumped into a tower filled with a packing material to enhance aeration and slow the water movement. Air is then blown into the base of the tower as the water falls through the tower, and the volatile compounds are stripped and entrained to the atmosphere (Figure 2). This technology requires that the contaminant be highly volatile (such as halogenated solvents or light hydrocarbon fuels) and can be fairly cost effective. The release of the contaminant into the atmosphere is being discouraged in some areas, and a "polishing" treatment is often required to capture the airborne contaminants in a secondary treatment process (such as carbon adsorption). This technology is effective for solvent and some fuels and is readily available if a permit to discharge to the atmosphere is obtainable.

Carbon Bed Adsorption

Carbon bed adsorption is a well-established technology that is used to treat contaminated groundwater (Figure 3). The contaminant becomes adsorbed to the carbon, which is most efficient in removing weakly polar molecules. The concentration of the contaminated influent directly affects the retention time of carbon adsorption before all the absorption sites are filled and breakthrough of the contaminants occurs. The spent carbon is then replaced with unused carbon, and the adsorption process can continue. Carbon is versatile and usable for many organic contaminants, but it can be costly. The quantity of carbon must be calculated based on the type of contami-

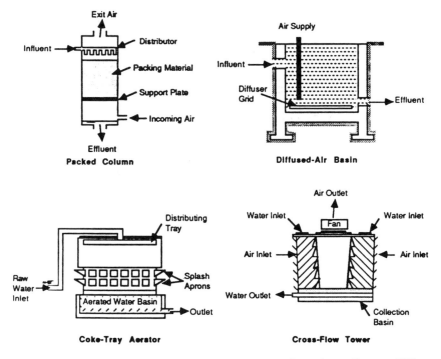

Figure 2. Air-stripping equipment and possible configurations. *Source:* EPA (1985).

nant, contaminant concentrations, and size of the container. Ultimately, the contaminant will break through as the available adsorption capacity of the carbon is used. Costs can become high, depending on how often carbon must be regenerated or replaced.

Incineration (Thermal Destruction)

Incineration involves thermal combustion of the material and, theoretically, rendering the contaminant into harmless by-products. Incinerators burn the material at high temperatures (1300–3000°F), and the combustion residence time varies with the type of contaminant. Combustion end-products release water and carbon dioxide through a permitted air discharge stack. Monitoring devices are placed in the discharge stack to ascertain if any contaminants or combustion by-products are leaving without complete combustion. Ash resulting from noncombustion is collected and landfilled or otherwise disposed. Currently, this technology is being used more widely for con-

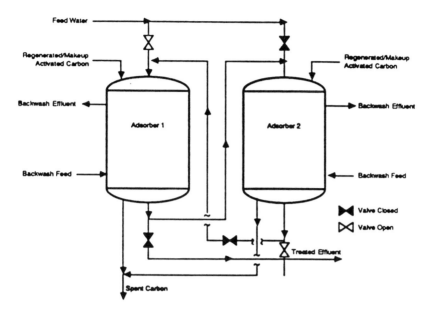

Figure 3. Flow diagram for a two-vessel granular carbon absorption system. *Source:* EPA (1979).

taminant treatment since current and future federal hazardous waste landfill regulations will not allow land disposal of some hazardous wastes, especially liquids. Incineration is also used for domestic refuse and agricultural waste destruction, and electric power production. This technology can be cost effective and is contaminant specific. Incineration technology utilizes rotary kiln, liquid injection, multiple hearth, or fluidized beds (Table 1).

Bioremediation

Numerous biologic treatment techniques have been developed for use in sewage treatment technology. Microbacteriological activity uses the contaminant as a food source, which renders it "nontoxic," producing "harmless" by-products. Aboveground treatment may involve aeration or enhanced biologic action in surface containers or tanks. Biologic remediation also may be used on excavated soil spread onto the land surface and supplied with nutrients and oxygen for microfauna to metabolize the contaminants. The efficiency of oxygen delivery to the microbiota usually governs the effectiveness of the bioremediation. Bioremediation has been used for treating excavated

Table 1. Summary of Commonly Used Incineration Technologies

Type	Process Principle	Application	Combustion Temperature	Residence Time
Rotary kiln	Waste is burned in a rotating refractory cylinder.	Any combustible solid, liquid, or gas	1500 to 3000°F	Seconds for gases; hours for liquids and solids
Single chamber/ liquid injection	Wastes are atomized with high-pressure air or steam and burned in suspension.	Liquids and slurries that can be pumped	1300 to 3000°F	0.1 to 1 second
Multiple hearth	Wastes descend through several grates to be burned in increasingly hotter combustion zones.	Sludges and granulated solid wastes	1400 to 1800°F	Up to several hours
Fluidized bed	Waste is injected into an agitated bed of heated inert particles. Heat is efficiently transferred to the wastes during combustion.	Organic liquids, gases, and granular or well processed solids	1400 to 1600°F	Seconds for gases and liquids; minutes for solids

Source: EPA (1985).

soils containing petroleum fuel contamination and has been used successfully on soil volumes ranging from hundreds to thousands of yards.

Subsurface bioremediation treatment involves recirculating contaminated water into the groundwater, which contains an oxygen (such as peroxide) and nutrients. This delivers nutrients to allow biologic action to start on contaminants, "cleaning" the aquifer matrix. The technology was pioneered by R. L. Raymond and Virginia Jamison during the 1960s. The groundwater is pumped by extraction wells, moving the treated waste through the contaminated zone and allowing biologic respiration of the indigenous microfauna to further metabolize remaining contaminants. The contaminant plume must not move offsite, and the site geology must be acceptable (sandy without abundant disseminated clay layers, which inhibit water movement). Nutrient delivery to the aquifer must be balanced so that waste products from the microbiota do not clog the porosity. This bioremedial approach is most suited to hydrocarbon fuel cleanup and has been used for other aliphatic compounds. Research is ongoing to ascertain the effectiveness of this technology on pesticides and other contaminants.

Soil Vapor Venting

Soil vapor venting involves venting the vapors of volatile contaminants from unsaturated subsurface soils in the vadose zone. An air suction is placed on extraction wells placed in the contaminated vadose zone area, and the vapors are removed and treated prior to atmospheric discharge (Figure 4 and Tables 2 and 3). This technology has been used for methane and vapor control along the perimeters of landfills for years. Although not a "groundwater" cleanup technology, it can remediate the vadose zone overlying the groundwater by removing residual saturation of contaminants that might contaminate the groundwater. Surface aeration of soil contaminated by petroleum fuels has been used in California and other states for remediation of soils. The soil is spread into thin (6-in.) layers and allowed to aerate daily, and then disced to expose unaerated soil. This passive technique is a relatively inexpensive option for soil cleanup.

Subsurface active vapor venting involves forced air movement through the affected area by a series of air injections into the extraction wells. Design data needed for these systems include knowledge of

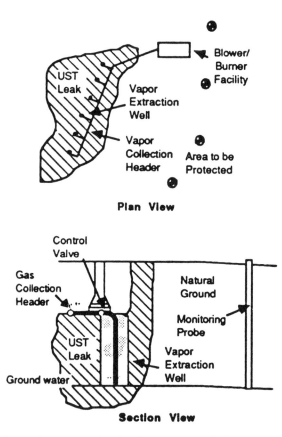

Plan View

Section View

Figure 4. General components of active vapor control systems and possible extraction well array. *Source:* EPA (1985).

the extent of vadose contamination and physical soil properties. Once the extent and estimated volume of contaminant are known, the vapor venting wells are installed, connected to blowers and vacuum equipment, and started. The contaminants are vented to the permitted discharge, with or without a polishing unit. During the venting, contaminant concentrations are measured to ascertain the effectiveness of the treatment. API (1989) and Johnson et al. (1989) have prepared papers on the design considerations for a soil venting system.

Table 2. Materials and Equipment for Active Vapor-Control Systems

Item	Materials	Installation
Well drilling		Auger, caisson, or bucket drill rig
Well piping	2- to 6-in. PVC, schedule 40 to 80, perforated and solid-wall	Crane for deep wells, backhoe for shallow wells
Well backfill	1-in. washed crushed stone or river gravel	Placed slowly by hand
Header piping	3-in. or greater (depending on flow/pressure requirements) PVC, polyethylene, or fiberglass (resistant to chemical attack)	Conventional trench excavating equipment, specialized jointing equipment for some pipe materials
Valves	Compatible with pipe size; gate, ball, or butterfly type; PVC or other chemical-resistant material	Jointing similar to piping materials
Vacuum blower	Material or coating that resists chemical attack; size varies with flow/pressure requirements	Foundation and installation per manufacturer recommendations
Safety devices	Specific items manufactured for use at refineries, sewage digestors, etc.	Installed with piping
Vent stacks	Any corrosion-resistant pipe of adequate size and strength; may require support	Same as header piping

Source: Data from EPA (1985).

Table 3. 1986 Unit Costs for Components of Active Vapor-Control Systems

Item	Range of Unit Costs ($)
Vapor-extraction well (drilling, stone, piping, etc.), in place	50 to 80/vertical foot
Well connection lateral (10-ft piping valve, excavation, fittings, etc.), in place	1,000 to 1,550 each
Vapor-collection header (piping, excavation, fittings, etc.), in place	20 to 105/linear foot
Blower facility (blower(s), safety devices, valves, foundation, piping, fencing, electrical components, and service connection), in place	50,000 to 105,000 (total)
Monitoring probe (drilling, pipe, fittings, backfill, etc.), in place	10 to 15/vertical foot
Operation and maintenance	5,000 to 20,500/year
Monitoring with portable meter	10 to 15 each visit

Source: Data from EPA (1987).

INTERIM CLEANUP ACTION

Interim cleanup steps may begin either before all the site investigation is completed or while some of the final cleanup plans are negotiated—such as the method of contaminant removal. Interim cleanup could involve any cleanup technology for limited vadose or saturated zone remediation and may precede the formulation of the remedial action plan in order to stabilize the contamination and prevent its spread. Interim cleanup commonly involves steps for temporary remediation, and the supervising agency will direct the initial cleanup action. Ultimately, the interim action will be viewed in the greater context of the entire project and integrated into final remediation plans.

For example, separate-phase (immiscible) product recovery is a typical interim action where lighter-than-water contaminants are physically removed from near the water surface. Numerous recovery technologies are available for use in small- or large-diameter wells. The removal of separate-phase product ameliorates the contaminant migration threat and dissolution into groundwater. Recovery may be done either by hand bailing or automated pumping systems using existing site monitoring wells, or wells which were installed to reveal presence of product. Materials and equipment must often be explosion proof, and the recovered product is temporarily stored onsite until it can be properly disposed. This is a relatively inexpensive and rapid remedial technique and is often considered as an interim remedial action by directly removing contaminant.

FINAL SITE REMEDIAL ACTION AND CLOSURE PLAN

When starting a site cleanup, a plan or orders are sent by the regulating agency to the individual or company stating when it should be done and what residual concentrations are allowable in the soil or water. Final allowable concentrations are usually based on toxicological threat to humans (including cancer risk) and the available pathways for the contaminant movement. The agency will require final postcleanup sampling and testing to check the effectiveness of cleanup and ensure that the allowable cleanup standard has been met. The lead agency will usually not issue a clean bill of health for any site and reserves the right to reopen the case if future information suggests it is necessary. This can have significant impact upon the future use of the land, or transferring land ownership (e.g., bank loans, rezoning) prior to, during, and following site remediation.

At some point a remedial action plan is prepared which summarizes the site investigation data and engineering review for site cleanup. This plan will outline the nature of the problem, all historic data, options for cleanup, cost analysis, and the selected cleanup option. Cleanup options include the BAT evaluation and, if warranted, discussion and use of experimental technologies. All rationales for data interpretation and selection factors are included for the review of agencies and the public. Postcleanup reporting will be required to ascertain the effectiveness of the remediation.

Safety issues have been strengthened in cleanup regulations that have been passed recently and must be closely followed. The preparation of the safety plan is beyond the scope of this text. However, all remedial plans must include a safety plan that meets the site requirements and all appropriate regulations. Many of these regulations deal with right-to-know laws and worker safety. Both are of paramount importance during any soil or groundwater contamination cleanup and should be rigorously followed to avoid future legal entanglements.

Example of State Agency Oversight in a Site Cleanup

The State of California (and other states) currently has agencies that address contamination and cleanup. Certain agencies take the lead on the project, and others will comment or consider other parts of a project as it relates to their sphere of influence. Ultimately, several governmental agencies may become involved. An example of state agencies that might review a project and have input in California include the Department of Health Services (DOHS), which deals with soil and surface contamination to remove hazardous threat to humans, and the Regional Water Quality Control Board (RWQCB), which deals with surface and aquifer water protection and restoration and the protection of beneficial uses.

Different agencies may participate at selected stages of remediation review and approval. The time for review, comment, and negotiation can increase with plan complexity and required cooperation among involved parties. The state may elect to give some of the regulatory oversight power to local agencies, water districts, or county agencies—commonly the local health agencies or hazardous management agencies—to assist in cleanup and site monitoring. The state will often provide information to assist the interested parties. Such information may include the following:

site assessment manual, issued by the state as a minimum standard for investigations

cleanup methodology manual, which sets rationale for establishing cleanup levels

leaking underground fuel tanks (LUFT) manual, an evolving manual specifically designed to assist consultants with leaking tanks

policy or clarification letters (or other communications) issued periodically to explain evolving positions or changing rationale

local agency ordinances, codes, and rules, which may dictate, or in some cases supersede, state policy

negotiation between the interested agencies, cities, or other groups, and potentially responsible parties and consultants

Finalizing the Cleanup Plan

Typically, the following steps are negotiated and resolved prior to starting the final cleanup. Any number of additional steps could be required by site conditions or regulatory order. These steps represent typical parts of the remedial action plan, as well as contractual or budget considerations.

Final demarkation of the area to be remediated, the so-called zero contamination line. Failure to locate the edge of cleanup, and agree to that boundary with the regulators, may result in additional investigation and cleanup. Additional hydrogeology and engineering costs may arise following the initial site investigation, to collect more information if the assessment is incomplete.

Acceptance of the cleanup plan by the regulating agencies (with public review if required). This will involve complete negotiation of contamination extent, cleanup methods, cleanup costs, and selected cleanup technology and related engineering design. Resolving these issues may take years and may require substantial changes in the technical approach (even possibly new investigations).

Securing the needed permits from the appropriate agencies. This may include excavation, equipment and well construction, waste discharge methodologies, and permits.

Preparation and review of the site safety plan and personnel medical monitoring as required.

Determination of the costs of cleanup. This varies depending on the remediation method and can include budgets for cleanup, maintenance, monitoring, and site closure. Cleanup costs are typically high, easily reaching the tens or hundreds of thousands of dollars (not including future monitoring or other efforts). Usually a cost analysis is included in a cleanup plan when the most cost-effective remedial option is determined. Cost may change due to inflation, additional equipment needs, additional review by regulators, effectiveness of the initial cleanup, or changes in subsurface conditions not anticipated in the site investigation.

Selection of the cleanup contractor, usually by competitive bid. The bid proposal would include the cleanup plan and documentation so that the contractor will perform the work in a cost-effective manner and negotiate additional funds for unforeseen contingencies or problems.

Finalization of any physical plant or containment facilities, including design engineering and drawings, work timetables, safety plan, and other site procedural documents.

Setting up the decontamination area for field work, or constructing and installing the needed treatment, monitoring, and safety equipment (including the calibrating and adjusting equipment, long-term maintenance, etc.).

Verification monitoring or sampling to evaluate the system's effectiveness for site cleanup. This would include interval reporting (usually quarterly) following system startup and preparing reports for regulatory oversight.

An example of a remedial action/closure plan format and its components follows:

1. Background history
   - cause and location of contaminant release
   - how release was detected
   - estimate of duration and volume of release
   - type of leak detection system installed at site
2. Site characterization
   - subsurface exploration and soil sampling methods
   - groundwater monitoring well design
   - groundwater sampling methods, water level measurements
   - sampling protocol (e.g., analytical chemical lab, chain-of-custody documentation, sampling preservation)
   - regulatory requirements for analyses performed
   - methods used to detect and measure separate-phase product
3. Extent of soil and groundwater pollution
   - vertical and lateral extent of subsurface contamination
   - number and location of exploratory borings
   - number and location of monitoring wells
   - definition of separation-phase product
   - definition of dissolved contaminants
4. Hydrogeology
   - subsurface lithologies, primary and secondary permeability
   - aquifer characteristics
   - aquifer and aquitard relationships
   - groundwater flow direction and gradient
   - seasonal and diurnal groundwater elevation changes
   - geologic cross sections
   - preferred contaminant pathways
   - permeability characteristics of the vadose zone
5. Beneficial uses of groundwater
   - existing and future groundwater uses
   - regional basin plan requirements
   - potential receptor/risk assessment
6. Remedial action
   - interim remedial actions used
   - development of remedial alternatives
   - screening remedial alternatives/technology/engineering

   • rationale for selected remedial action
   • soil remediation method (e.g., excavation, vapor venting)
   • groundwater remediation method
   • potential/existing impact of remedial action(s)
7. Effectiveness of remediation
   • consistency of cleanup levels with federal/state guidelines
   • verification monitoring program
   • potential impacts from residual contamination
8. Site closure
   • closure plan reviewed by lead agency
   • verification monitoring data reviewed
   • "sign-off" (no further work required)

EXAMPLE OF BIOREMEDIATION GROUNDWATER CLEANUP

A spill of about 1000 gal of gasoline occurred at a service station site in the central coast region of California (Figure 5). The portion of the spill retained in the vadose zone was excavated when the station was rebuilt and subsurface storage tanks replaced. An investigation was conducted to finish groundwater contamination definition. Initially, air stripping was proposed for groundwater remediation; however, this was vetoed by the local air quality board. Consequently, a closed-loop subsurface bioremediation approach was suggested. The key to successful bioremediation is a porous and permeable aquifer, which allows water to be continuously withdrawn and reinjected in order to deliver water and nutrients to the contaminated area. Site geology proved to be dune sand and marine terrace deposits, which are porous and permeable and suitable for bioremediation (Figure 6).

Problem Approach

Since the California Regional Water Quality Control Board (RWQCB) needed to approve subsurface injection of contaminated water, additional aquifer delineation was required to show that it was separate from the regional sole-source aquifer, providing the only source of drinking water (Safe Drinking Water Act of 1974). Dissolved contaminant levels ranged from 10 to 30 ppm of total petroleum hydrocarbons and 100 to 200 ppb benzene. Benzene concentration cleanup levels (the most sensitive contaminant) were negotiated

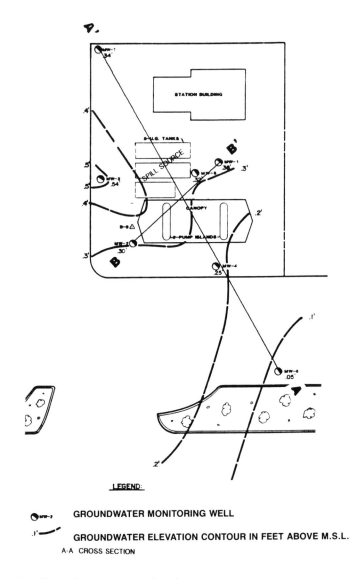

Figure 5. Pre-test groundwater elevation map.

to 7 ppb because the contaminated water was located in the perched aquifer system.

Additional investigations showed that the groundwater occurrence beneath the site was only semiperched and discontinuously in connec-

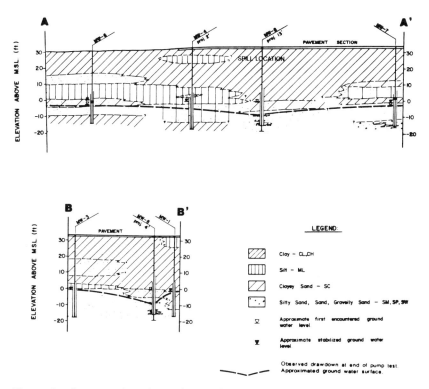

Figure 6.   Cross section of complex stratigraphy beneath bioremediation site.

tion with the underlying sole-source aquifer (see Figure 6). Soil samples were also collected to model growth of the indigenous microfauna that would be used for the cleanup and to estimate the quantity of gasoline in the aquifer porosity.

An aquifer performance test was conducted to ascertain the optimum yield of the extraction well, observe the extent of the cone of influence of the well, and use the data to estimate nutrient-charged water injection rates. A step and constant discharge test was performed, which determined that a well yield of 13 gpm would create a cone sufficient to capture the contaminant plume (Figure 7). Injection rates and resulting mounds would have to stay within the cone and not disrupt the capture symmetry. The extracted groundwater would then be charged with the required nutrients and oxygen (as peroxide), reinjected around the periphery of the cone, and returned to the extraction well. The system was plumbed together, and water

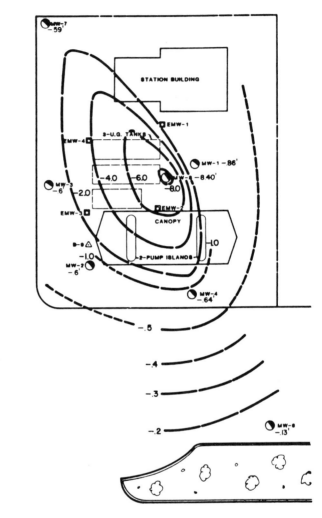

Figure 7. Groundwater drawdown map at 24 hours; cone assymetry due to boundary between pumping well and MW-1.

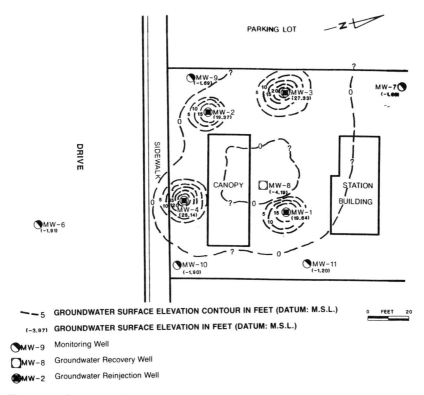

Figure 8. Recirculating bioremediation groundwater map; injection mounds are captured by the central extraction well.

recirculated to begin the remediation. The original cleanup time was estimated at 18 months.

System modifications were required during startup to balance flow and limit siltation of the injection wells. Both filters and well redevelopment were needed to remove silt to keep the flow at the design rates. After 6 months of operations, the contamination levels began to decline, and monthly declines were observed thereafter. The system was shut off during a regional aquifer water level decline below the base of the monitoring and extraction wells. Once groundwater had recovered in the fall, the system was restarted and adjusted. A typical pattern of injection and extraction is presented in Figure 8. The site owner decided to accelerate the pace of nutrient delivery to speed up the cleanup. After one year of operation, the concentration of all contaminants had declined to "none detected," and the system was turned off. Monthly monitoring was continued for one year to assure that the cleanup was complete.

Summary

Bioremediation was effective for cleanup of petroleum hydrocarbons contained in the aquifer at this site. The key to success is the ability to recirculate groundwater in a pump and injection system. The nutrient and oxygen delivery rates must be matched to the indigenous microfauna ability to metabolize the contaminants. Unless water can be moved through the strata, the cleanup can be inefficient. The ability to bioremediate at this site allowed a cleanup of dissolved contaminants and contaminants in the aquifer matrix, which protects the underlying sole-source aquifer.

VADOSE SOIL CLEANUP REMEDIATION PLAN DEVELOPMENT FOR A 2,4-DICHLOROPHENOXY ACETIC ACID SPILL (After Blunt, 1988)

Problem

In 1978 a large surface spill of 2,4-dichlorophenoxy acetic acid (2,4-D) occurred at a bulk transfer and product formulation facility in central California (CCMF). Several phases of subsurface exploration (36 borings) defined the liquid 2,4-D spill to be in the shallow vadose zone sandy and gravelly sediments at depths of 10 to 20 ft (Figure 9). The investigation and definition phases took over 7 years to complete. The exploratory soil boring and sample analyses program ultimately defined the extent and depth of penetration of the contaminant. In order to clean up the spill, a remedial action plan was needed to address the most cost-effective method of 2,4-D removal that would minimize exposing the surrounding residential areas (Figure 10).

Exposure Risk and Contaminant Pathway Identification

When the pesticide was spilled, it became a waste that state agencies regulated at 100 ppm for waste identification. 2,4-D is known to have toxic effects upon humans at acute, subchronic, and chronic levels. Stomach ulcers and animal birth defects may be caused by 2,4-D, and

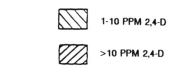

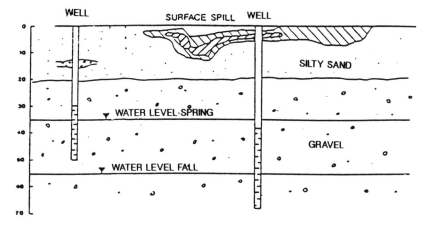

Figure 9. Geologic cross-section of 2,4,-D contamination in vadose zone.

it is a potential human carcinogen. Consequently, the health risks of this compound need to determine the likelihood of injury from human exposure pathways. A judgment of the margin of safety had to be determined from the possible exposure routes. Since human exposure routes include dermal, inhalation, and ingestion, an evaluation of possible human exposure concentration and cancer risk was required to assess tolerable carcinogenic risk.

Pathways of exposure from 2,4-D were reviewed given the chemical characteristics of the compound. The compound has a low vapor pressure, and movement by vapor diffusion would be negligible (Jury, Specer, and Farmer, 1983). 2,4-D is unlikely to be transported as vapor, but could be transported as dust. It could migrate by direct application if applied as a liquid, leaching into the soil. Once in the soil, the mechanisms of movement would be by mass flow, liquefied diffusion, and gaseous diffusion (Hern and Melancon, 1986). Once in the vadose zone, 2,4-D may reach the groundwater by leaching and be drawn into wells supplying water to the public. The soil type and concentration of organic matter in the soil influence both mobility

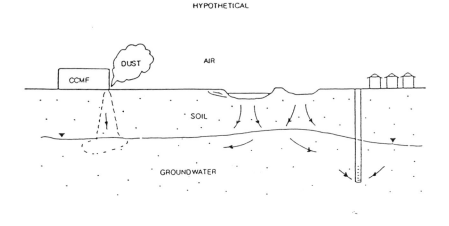

Figure 10. Potential migration and exposure routes of pesticide. *Source*: Blunt (1988).

and persistence of the pesticide. Although pesticides may leach downward in soils, they move slowly and short distances (Norris, 1966).

Developing the Remedial Action Cleanup Alternatives

Once the vadose contamination had been defined, remedial alternatives were needed for site cleanup. Five alternatives were reviewed for site remediation:

1. monitoring only
2. enhanced bioremediation and bioremediation
3. excavation to 10 ppm 2,4-D
4. excavation to 1 ppm 2,4-D
5. excavation to < 1 ppm 2,4-D

These alternatives were explored in terms of lessening the threat of human exposure and removing the contaminant. Safety factors, duration of the remediation, and cost for each option were also calculated (Table 4). The cost for exposure and safety are weighted against the cost for cleanup.

The "monitor only" alternative is the least costly and would not

Table 4. Summary of Parameters Used to Evaluate Cleanup Alternatives

Alternative	2,4-D Concentration	Material[a]	Dose[b]	Cancer Risk[c]	Safety Factor[d]	Cleanup Cost[e]
1	1,600 ppm	Soil	2.3×10^{-3}	4.5×10^{-10}	1.3	$ 72,400
	110 ppb	Water	3.1×10^{-3}	6.1×10^{-10}	0.9	
2	100 ppm	Soil	1.4×10^{-4}	2.8×10^{-11}	21	$ 1,158,000 to
	110 ppb	Water	3.1×10^{-3}	6.1×10^{-10}	1.1	$ 720,000
3	10 ppm	Soil	1.4×10^{-5}	2.8×10^{-12}	210	$ 470,000
	10 ppb	Water	2.9×10^{-4}	5.7×10^{-11}	10	
4	1 ppm	Soil	1.4×10^{-6}	2.8×10^{-13}	2100	$ 600,000
	1 ppb	Water	2.9×10^{-5}	5.7×10^{-12}	100	
5	0 (<1) ppm	Soil	$<1.4 \times 10^{-6}$	$<2.8 \times 10^{-13}$	>2100	$ 1,210,000
	0 (<1) ppb	Water	$<2.9 \times 10^{-5}$	$<5.7 \times 10^{-12}$	>100	

Source: Blunt (1988).
[a]Ingestion exposure.
[b]Calculated for a 70-kg adult as mg/kg body weight/day by oral ingestion.
[c]Calculation based on a *pica* of a 1-day maximum exposure.
[d]The ratio of the MCL dose to the site calculated dose (0.1 mg/L × 2L/70 kg)/Theoretical site dose 2,4-D).
[e]All costs assume a cleanup and monitoring program of 10 years at 1988 dollars.

remove any contaminant. Blunt (1988) calculated that the time required for pesticide to naturally degrade to low levels is long compared to other alternatives, and provides a safety factor of 1.3. The bioremediation alternative would use a technology that is not proven for this contaminant and would involve additional study to determine effectiveness. Study of the bioremediation process and models indicated that the pesticide may migrate, and the ultimate cleanup effectiveness may not be definable.

The three excavation alternatives would remove the pesticide. The calculated safety factors increase and exposure risks decrease with the increased excavation; hence, a cost for exposure risk must be considered. Safety factors rise as the level of excavation effort rises, until all of the contaminant is removed.

Remedial Action Selection

A comparison of remedial alternatives suggested that the third excavation alternative most quickly removes the problem and renders the site safe from long-term exposure. Areas of elevated concentrations of pesticide would be excavated and transported to a hazardous waste landfill. The small quantity of pesticide remaining would be covered by an impermeable cap to preclude infiltrating water from leaching the pesticide deeper. The reduced quantity of contaminant and greatly lessened ability for vertical migration produced the greatest benefit for the cost and effort expended. The cost for benefit (safety factor) was the negotiable issue, and the third excavation alternative proved the best compromise to all interested parties.

SUMMARY

Site remediation will be a compromise based on the applicable regulations, hydrogeologic data, data interpretation, the final cleanup concentrations of contaminants, negotiation of the remedial action plan alternatives, the cost, and remediation plan execution. Consultants and remediation contractors cannot completely remove all of the contaminant from the site subsurface. The ability to estimate the safety risk and cost-benefit ratio of the remediation effort should be used. A remedial action plan based on accurate subsurface information, applied within regulatory guidelines and within the ability of current technology and available budget, is the best compromise. Given the myriad legal, scientific, and engineering problems faced in even small-scale cleanups, striving for a unobtainable utopian cleanup goal is counterproductive. A maximum effort with the most innovative and efficient use of the technology and available funds is the best remedial approach.

References

Abdul, A. S., T. L. Gibson, and D. N. Rai. 1990. "Laboratory Studies of the Flow of Some Organic Solvents and Their Aqueous Solutions through Bentonite and Kaolin Clays," *Groundwater* 28:524–533.

American Petroleum Institute. 1983. "Groundwater Monitoring and Bias," API Pub. No. 5367, June, Washington, DC.

American Petroleum Institute. 1989. "A Guide to the Assessment and Remediation of Underground Petroleum Releases," API Pub. No. 1628, August, Washington, DC.

American Society for Testing and Materials (ASTM). 1988. *Annual Book of ASTM Standards,* Section 4, Construction, Volume 04.08, Soil and Rock, Building Stones, Geotextiles, Methods D 420-87, D 653-87,D 2487-85, D 2488-84 (Philadelphia: ASTM).

Anderson, M.P. 1984. "Movement of Contaminants in Groundwater: Groundwater Transport, Advection, and Dispersion," in *Studies in Geophysics*, (Washington, DC: National Academy Press), pp. 37–45.

Association of Engineering Geologists. 1981 (revised 1985). "Professional Practice Guidelines," Sections 1–9, Special Publications, AEG, Lawrence, KS.

Back, W., J. S. Rosenshein, and P. R. Seaber. 1988. *Hydrogeology: The Geology of North America,* Vol. 0-2 (Boulder, CO: Geological Society of America).

Barcelona, M.J., and J.A. Helfrich. 1986. "Well Construction and Purging Effects on Groundwater," *Environ. Sci. Technol.* 20:1179–1184.

Barker, J.F., G.C. Patrick, and D. Major. 1987. "Natural Attenuation of Aromatic Hydrocarbons in a Shallow Sand Aquifer," *Groundwater Monitoring Review* 8:64–71.

Behnke, J., C. M. Palmer, D. Peterson, and J. L. Peterson. 1990. *Groundwater Contamination and Field Investigation Methods; Workshop Notebook* (Chico, CA: California State University).

Bentall, R., Ed. 1963. "Shortcuts and Specific Problems in Aquifer Tests," U.S. Geological Survey Water Supply Paper 1545-C.

Birkeland, P. W. 1984. *Soils and Geomorphology* (New York: Oxford University Press).

Blunt, D.J. 1988. "Delineation and Remediation Action Planning — A Spill of the Pesticides 2,4-D and 2,4,5-T," in *Proc. Hazardous Materials Conference*, Association of Bay Area Governments, Anaheim, CA, pp. 376–392.

Boulton, N. S. 1963. "Analysis of Data from Non-Equilibrium Pumping Tests Allowing for Delayed Yield Storage," *Proc. Inst. Civ. Engr.* 26:469–482.

Bouwer, H., and H. Rice. 1976. "A Slug Test for Determining Hydraulic Conductivity of Unconfined Aquifers with Completely or Partially Penetrating Wells," *Water Resour. Res.* 12(3):423–428.

Cherry, J.A., R.W. Gilliam, and J.F. Baker. 1984. "Contaminants in Groundwater: Chemical Processes," in *Studies in Groundwater* (Washington, DC: National Academy Press), pp. 46–68.

Cooper, H. H., Jr., J. D. Bredehoeft, and I. S. Papadopulas. 1967. "Response of a Finite-Diameter Well to an Instantaneous Charge of Water," *Water Resour. Res.* 3(1):263–269.

Cooper, H. S., Jr., and C. E. Jacob. 1946. "A Generalized Method for Evaluating Formation Constants and Summarizing Well-Field History," *Am. Geophys. Union Trans.* 27(4):526–534.

Davis, S. H. 1987. "What is Hydrogeology?" *Groundwater* 25:2–3.

Davis, S.N., and R.J.M. DeWiest. 1966. *Hydrogeology* (New York, NY: John Wiley & Sons).

Department of Agriculture (U.S.). 1979. *Field Manual for Research in Agricultural Hydrology,* Ag. Handbook No. 224.

Department of Interior (U.S.). 1981. *Ground-Water Manual — A Water Resources Technical Publication,* Wiley-Interscience Publication.

Driscoll, F. G. 1986. *Groundwater and Wells,* 2nd ed. (St. Paul, MN: Johnson Filtration Systems, H. M. Smyth Co.).

Dunlap, L. E. 1985. "Sampling for Trace Level Dissolved Hydrocarbons from Recovery Wells Rather Than Observation Wells," *Proceedings of the Petroleum Hydrocarbons and Organic Chemicals in Ground Water—Prevention, Detection, and Restoration,* Houston, TX: National Water Well Association, pp. 223–235.

Dupuit, J. 1863. *Etudies theoriques et pratiques sur le mouvement des eaux dans les canaux decouverts et a travers les terrains permeables,* 2nd ed. (Paris: Dunot).

Elliott, J. 1987–1991. *The Toxics Program Matrix (California, Florida, Illinois, New Jersey, Ohio, Pennsylvania, Texas)* (Toronto: Specialty Publishers).

Environmental Protection Agency (U.S.). 1984a. "Permit Guidance Document on Unsaturated Zone Monitoring for Hazardous Land Treatment Units," EPA/530–800–84–016.

Environmental Protection Agency (U.S.). 1984b. "Soil Properties Classification and Hydraulic Conductivity Testing," EPA SW-925.

Environmental Protection Agency (U.S.). 1985a. "Practical Guide for Ground Water Sampling," EPA/600/2–85/10.

Environmental Protection Agency (U.S.). 1985b. "Protection of Public Water Supplies from Ground-Water Contamination," Seminar Publication, EPA Center for Environmental Research, Cincinnati, OH, EPA/625/4–85/016.

Environmental Protection Agency (U.S.). 1986a. "EPA RCRA Ground-Water Monitoring Technical Enforcement Guidance Document."

Environmental Protection Agency (U.S.). 1986b. "Test Methods for Evaluating Solid Waste Physical/Chemical Methods," EPA SW-846, Vols. 1A, 1B, and 1C.

Environmental Protection Agency (U.S.). 1987a. "DRASTIC: A Standardized System for Evaluating Ground Water Pollution Potential Using Hydrogeologic Settings," EPA-600/2–89–035.

Environmental Protection Agency (U.S.). 1987b. "Underground Storage Tank Corrective Technologies," Hazardous Waste Engineering Research Laboratory, EPA/625/6–87–015.

Environmental Protection Agency (U.S.). 1989. "Transport and Fate of Contaminants in the Subsurface," Center for Environmental Research Information, Cincinnati, OH, and Robert S. Kerr Environmental Laboratory, Ada, OK.

Ferris, J. G., and D. B. Knowles. 1954. "The Slug-Injection Test for Estimating the Coefficient of Transmissivity of an Aquifer," U.S. Geological Survey Water Supply Paper 1536-J.

Fetter, C. W., Jr. 1988. *Applied Hydrology* (New York: C. E. Merrill Publishing Co.).

Freeze, R. A., and J. A. Cherry. 1979. *Groundwater* (Englewood Cliffs, NJ: Prentice-Hall).

Gear, B. B., and J. P. Connelley. 1985. "Guidelines for Monitoring Well Installation," *Fifth National Symposium and Exposition on Aquifer Restoration and Ground Water Monitoring,* pp. 83-104.

Geraghty, J.J. and D.W. Miller. 1985. "Fundamentals of Groundwater Contamination," Short Course Notebook: Geraghty and Miller, Inc., Syosset, NY.

Gibs, J., and T.E. Imbrigotta. 1990. "Well Purging Criteria for Sampling Purgable Organic Compounds," *Groundwater* 28:68-78.

Glass, R., E. Steenhur, and J. Parlange. 1988. "Wetting Front Instability as a Rapid and Far-Reaching Hydrologic Process in the Vadose Zone," *J. Contam. Hydrol.* 3:207-226.

Gymer, R.G. 1973. *Chemistry: An Ecological Approach* (New York, NY: Harper and Row).

Hackett, G. 1987. "Drilling and Constructing Monitoring Wells with Hollowstem Augers: Part I, Drilling Considerations," *Groundwater Monitoring Rev.* 7:51-62.

Hantush, M. S. 1956. "Analysis of Data from Pumping Tests in Leaky Aquifers," *Am. Geophys. Union Trans.* 37:702-714.

Hantush, M. S. 1960. "Modification of the Theory of Leaky Aquifers," *J. Geophys. Res.* 65:3713-3725.

Hantush, M. S. 1962. "Aquifer Tests on Partially Penetrating Wells," *Am. Soc. Civ. Engr. Trans.* 127(Part I):284-308.

Harrill, J. R. 1970. "Determining Transmissivity from Water-Level Recovery of a Step-Drawdown Test," U.S. Geological Survey Prof. Paper 700-C, pp. C212-C213.

Healy, B. 1989. "Monitoring Well Installation Misconceptions about Mud Rotary Drilling," *National Drilling Buyers Guide.*

Heath, R. C. 1982. "Basic Groundwater Hydrology," U.S. Geological Survey Water Supply Paper 2220.

Heath, R. C. 1984. "Ground-Water Regions of the United States," U.S. Geological Survey Water Supply Paper 2242.

Heath, R. C., and F. W. Trainer. 1981. *Introduction to Ground Water Hydrology* (Dublin, OH: Water Well Journal Publishing Co.).

Hem, J.D. 1970. "Study and Interpretation of the Chemical Characteristics of Natural Groundwater," U.S. Geological Survey Water Supply Paper 1473.

Hern, S. C., and S. M. Melancon. 1986. *Vadose Zone Modelling for Organic Pollutants* (Chelsea, MI: Lewis Publishers).

Hillel, D. 1980. *Fundamentals of Soil Physics* (New York: Academic Press).

Hodapp, D., J. Sagebiel, and S. Tester. 1989. "Introduction to Organic Chemistry for Hazardous Materials Management," short course, University of California, Davis.

Hughes, J.P., C.R. Sullivan, and R.M. Zinner. 1988. "Two Techniques for Determining Thickness in an Unconfined Sandy Aquifer," in *Proc. Petroleum Hydrocarbons Conference*, Houston, TX, National Water Well Association.

Jacob, C. E. 1963. "Correction of Drawdowns Caused by a Pumped Well Tapping Less Than a Full Thickness of an Aquifer," U.S. Geological Survey Water Supply Paper 1536-I, pp. 272–282.

Johnson, P.C., C.C. Stanley, M.W. Kemblowski, D.L. Byers, and J.P. Colthart. 1990. "A Practical Approach to the Design, Operation and Monitoring of an In-Situ Soil-Venting System," *Groundwater Monitoring Review* 10:159–178.

Jury, W. A., W. F. Specer, and W. J. Farmer. 1983. "Use of Model for Predicting Relative Volatility, Persistence, and Mobility of Pesticides and Other Trace Organics in Soil Systems," in *Hazardous Assessment of Chemicals,* Vol. 2 (New York: Academic Press).

Keys, W. S., and L. M. MacCary. 1971. "Application of Borehole Geophysics to Water Resource Investigations," Techniques of Water Resources Inv., U.S. Geological Survey, Chapter E1, Book 2.

Kruseman, G. P., and N. A. DeRidder. 1976. "Analysis and Evaluation of Pumping Test Data," ILRI-ISBN 90–70260–808, Bulletin No. 11.

LeRoy, L. W., D. O. LeRoy, and J. W. Raese. 1977. *Subsurface*

Geology, Petroleum, Mining, Construction (Golden, CO: Colorado School of Mines).

Loheman, S. W. 1979. "Ground-Water Hydraulics," U.S. Geological Survey Professional Paper 708.

Mackay, D. M., P. U. Roberts, and J. A. Cherry. 1985. "Transport of Organic Contaminants in Groundwater," *Environ. Sci. Technol.* 19:1–9.

Marbury, R. E., and M. E. Brazie. 1988. "Groundwater Monitoring in Tight Formations," in *Proceedings of the Second Annual Outdoor Action Conference on Aquifer Restoration, Groundwater Monitoring, and Geophysical Methods,* (Las Vegas) Vol. I, National Water Well Association, Dublin, OH.

Mathewson, C. W. 1979. *Engineering Geology* (Columbus, OH: Charles E. Merrill Co.).

Matthess, G. 1982. *The Properties of Groundwater* (New York, NY: John Wiley & Sons).

McAlony, T. A., and J.F. Barker. 1987. "Volatilization Losses of Organics During Ground Water Sampling from Low Permeability Materials," *Ground Water Monitoring Review* 7:63–68.

McCray, K. B. 1986. "Results of Survey of Monitoring Well Practices among Ground Water Professionals," *Ground Water Monitoring Review* 7:37–38

McCray, K. B. 1988. "Contractors Optimistic about Monitoring Business," *Water Well Journal, NWWA* 42:45–47.

Merry, W., and C. M. Palmer. 1985. "Installation and Performance of a Vadose Monitoring System," *Conference on Monitoring the Unsaturated (Vadose) Zone* (Denver: NWWA), pp. 107–125.

Miller, D. W., Ed. 1980. *Waste Disposal Effects on Ground Water* (Berkeley, CA: Premier Press), pp. 511–512.

Montgomery, J. H., and L. M. Welkom. 1989. *Ground Water Chemicals Desk Reference* (Chelsea, MI: Lewis Publishers).

Moore, J.W., and S. Ramamoothy. 1984. *Organic Chemicals in Natural Waters* (New York, NY: Springer-Verlag).

Morris, D. A., and A. I. Johnson. 1967. "Summary of Hydrologic and Physical Properties of Rock and Soil Materials, As Analyzed by the Hydrologic Laboratory of the U.S. Geological Survey 1948–1960," U.S. Geological Survey Water Supply Paper 1839-D.

Morrison, R. D. 1989. "Uncertainties Associated with the Transport and Sampling of Contaminants in the Vadose Zone," paper presented at the Association of Engineering Geologists meeting, Sacramento, CA, March 30.

Muskat, M. 1946. *The Flow of Homogeneous Fluids through Porous Media* (Ann Arbor, MI: J. W. Edwards).

National Water Well Association. 1988. "Introduction to Groundwater Geochemistry," short course, presented by the Association of Groundwater Scientists and Engineers, San Francisco.

Neuman, S. P. 1972. "Theory of Flow in Unconfined Aquifers Considering Delayed Response of the Water Table," *Water Resour. Res.* 8:1031–1045.

Neuman, S. P. 1975. "Analysis of Pumping Test Data from Anisotropic Unconfined Aquifers Considering Delayed Gravity Response," *Water Resour. Res.* 11(2):329–342.

Newsom, J. M. 1985. "Transport of Organic Compounds Dissolved in Ground Water," *Ground Water Monitoring Review* 3:41–48.

Nielson, D. M., and A. I. Johnson, Eds. 1990. "Ground Water and Vadose Zone Monitoring," American Standard Testing Methods Symposium, Albuquerque, NM. American Society for Testing and Materials.

Nielson, D. M., and G. L. Teates. 1985. "A Comparison of Sampling Mechanisms Available for Small-Diameter Ground Water Monitoring Wells," *Fifth National Symposium and Exposition on Aquifer Restoration and Ground Water Monitoring,* National Water Well Association, Dublin, OH, pp. 237–270.

Norris, L. A. 1966. "Degradation of 2,4-D and 2,4,5-T in Forest Litter," *J. For.* 64:475.

Palmer, C.M., and J. Elliott. 1988. "Now My Land is Contaminated: Whom Must I Tell and What Must I Do?" in *Proc. HazMat West Conference and Exposition*, Long Beach, CA, pp. 537–539.

Patrick, R., E. Ford, and J. Quarles. 1987. "Federal Statutes Relevant to the Protection of Ground Water," in *Legal Issues in Groundwater Protection* (Philadelphia: American Law Institute, American Bar Association), pp. 6–45.

Peterec, L., and C. Modesitt. 1985. "Pumping from Multiple Wells Reduces Water Production Requirements: Recovery of Motor Vehicle Fuels, Long Island, NY," in *Proceedings of the Petroleum Hydrocarbons and Organic Chemicals in Ground Water—Prevention, Detection and Restoration,* National Water Well Association, Dublin, OH, pp. 358–373.

Reineck, H. E., and I. B. Singh. 1986. *Depositional Sedimentary Environments* (Berlin: Springer-Verlag).

Robbins, G. A., and M. M. Gemmell. 1985. "Factors Requiring Resolution in Installing Vadose Zone Monitoring Systems," *Fifth National Symposium and Exposition on Aquifer Restoration and Ground Water Monitoring,* National Water Well Association, Dublin, OH, pp. 184–196.

Rosenshein, J. S., J. B. Gonthier, and W. B. Allen. 1968. "Hydrologic Characteristics and Sustained Yield of Principal Ground-Water Units, Potowomut-Wickford Area, Rhode Island," U.S. Geological Survey Water Supply Paper 1775.

Santa Clara Valley Water District. 1989. "Investigation and Remediation at Fuel Leak Tanks—Guidelines for Investigation and Technical Report Preparation."

Sax, N. I., and R. J. Lewis. 1987. *Hawley's Condensed Chemical Dictionary* (New York: Van Nostrand Reinhold Co).

Schneider, W. J. 1970. "Hydrologic Implications of Solid-Waste Disposal," U.S. Geological Survey Circular 601-F.

Schmelling, S.G., and R.R. Ross. 1989. "Contaminant Transport in Fractured Media: Models for Decision Makers," EPA Superfund Issue Paper, EPA 540/4–89/004, August.

Schmidt, K.D. 1982. "How Representative Are Water Samples Collected from Wells?" in *Proc. Second National Symposium on Aquifer Restoration and Groundwater Monitoring*, Columbus, OH. (Dublin, OH: National Water Well Association).

Schwille, F. 1988. *Dense Chlorinated Solvents in Porous and Fractured Media,* English language ed. (Chelsea, MI: Lewis Publishers).

Skoog, D. A., and D. M. West. 1971. *Principles of Instrumental Analysis* (New York: Holt, Rinehart, and Winston).

Soil Conservation Service (U.S.). 1978. "Ground Water," Water Resources Publications, National Engineering Handbook, Section 18.

State of California. Title 23, Waters, Chapter 3, Subchapter 16—Underground Tank Regulations, Articles 1-10.

State of California. 1973. *Design Manual,* Department of Transportation (revised 1984 by University of California, Inst. Trans. Studies, Berkeley).

State of California. 1986. The California Site Mitigation Decision Tree Manual, Dept. of Health Services, Sections 1-9 with appendices, Sacramento, CA.

Streltsova, T. D. 1974. "Drawdown in a Compressible Unconfined Aquifer," *J. Hyd. Div., Proc. of ASCE* 100(11):1601-1616.

Sykes, A.L., R.A. McAllister, and J.B. Homolya. 1986. "Sorption of Organics by Monitoring Well Construction Materials," *Groundwater Monitoring Review* 2:31-49.

Theis, C. V. 1935. "The Relationship between the Lowering of the Piezometric Surface and the Rate and Duration on a Well Using Ground Water Storage," *Am. Geophys. Union Trans.* 16(2):519-524.

Todd, D. K. 1980. *Groundwater Hydrology,* 2nd ed. (New York: John Wiley and Sons).

Toth, J. 1984. "The Role of Regional Gravity Flow in the Chemical and Thermal Evolution of Groundwater," *First Canadian/American Conference on Hydrogeology, Banff, Canada,* National Water Well Association, Dublin, OH, pp. 3-39.

U. O. P. Johnson. 1975. *Ground Water and Wells* (Saint Paul, MN: U. O. P. Johnson).

University of Missouri, Rolla. 1981. Seminar for Drillers and Exploration Managers Note Set, December 14-16, Phoenix, AZ.

USEPA. 1986. Groundwater Monitoring Seminar Series, CER 1-87-8, Slide copies.

USEPA. 1987. Handbook: Groundwater, EPA 625/6-87-016, March.

Vishner, G. S. 1965. "Use of Vertical Profile in Environmental Reconstruction," *Bull. Am. Ass. Petrol. Geol.* 49:49-61.

Walton, W. C. 1962. "Selected Analytical Methods for Well and Aquifer Evaluation," Illinois State Water Survey Bulletin 49.

Walton, W. C. 1970. *Groundwater Resource Evaluation* (New York: McGraw-Hill Book Co.).

Welenco, Inc. 1985. Electric Logging, Welenco, Inc.

Williamson, D. A. Undated. "The Unified Rock Classification," U.S. Forest Service, Willamette National Forest, Eugene, OR.

Wilson, L.G. 1981. "Monitoring in the Vadose Zone, Part II," *Groundwater Monitoring Review* 2:31–49.

Wilson, J.L., S.H. Conrad, E. Hagan, W.R. Mason, and Peplinski. 1988. "The Pore Level Spacial Distribution and Saturation of Organic Liquids in Porous Media," in *Proc. of Petroleum Hydrocarbons Conf.*, Houston, TX, National Water Well Association, pp. 107–133.

Index